Hydraulic Extrusion Presses

Hydraulic
Extrusion Presses

Presses used for the production of tubes, solid rods, hollow sections, wire, and cable sheathing in nonferrous metals

By

Ernst Müller

Duisburg

With 189 Illustrations

Springer-Verlag

Berlin/Göttingen/Heidelberg

1961

ISBN 978-3-642-53025-8   ISBN 978-3-642-53023-4 (eBook)
DOI 10.1007/978-3-642-53023-4

Preface

The wide variety of hydraulic presses used for the manufacture of tubes, cable sheathing, rods, profiles, and wire in nonferrous metals has gone through an extraordinarily interesting development which commenced about 50 years ago after the advent of high-tensile hot working steels and has since adapted itself to the steadily growing demands of the metalworking industry. This development has especially been promoted in Germany where many of the old presses are still in use for the extrusion of rods and tubes in heavy metals, and lead, already decades ago, to the construction of hydraulic presses which rank among the largest equipment built for noncutting shaping of metals.

This book, therefore, draws in the main upon the German development, giving an account of the advances made in extrusion technique and press design and presenting to the reader the sum of the experience gathered.

A great number of very instructive works dealing with the technological aspects in the extrusion of metals and the results obtained in metallurgical investigations has been published. These, it is true, contain important reference material regarding the design of presses and press tooling, but very few of them give a detailed description of the development and calculation of the machines or the requirements to be met by presses and tooling, so that the present book is to fill a gap in the existing literature.

This volume not only brings a comparative description of presses, press drives, press controls, auxiliary equipment, and tooling, but also indicates details of design and calculation which are the result of practical experience. In order to limit the scope of this book, such auxiliary equipment as runout and cooling beds, extrusion pullers, material-handling equipment, stretchers, bar breakers, cable take-up stands, wire coilers, etc., has been excluded. Special presses as used for cold extrusion, extrusion of special alloys, etc., and steel extrusion presses will receive separate treatment in a later, enlarged edition.

In his work the author has been generously assisted by many companies at home and abroad whose names are mentioned in the text and to all of whom he extends his thanks. He is especially indebted to his associates Dipl.-Ing. A. HERTL and Dd. E. LOHMANN who translated the book. It is hoped that this volume will be accorded a good reception by the metalworking industry and that it will supply valuable information to all readers.

Duisburg, May 1961   E. Müller

Contents

Chapter I

PRESSES FOR THE MANUFACTURE OF LEAD PIPES AND WIRE

The making of seamless lead pipes has been known for a long time and may be traced back to the early days of constructing hydraulic presses. Up to the present time changes in the technique of extrusion have only been slight. The bore of the container and the mandrel rod, being disposed in the center line of the container, form a hollow cylindrical space which is filled with molten lead. The end of the container bore is closed by a die, leaving open an annulus around the mandrel. From the opposite end the lead is extruded – after solidification – by a hollow ram, carrying at its nose a ring-shaped dummy block and traveling over the mandrel; it issues from the annulus in the form of a pipe.

The casting temperature of the lead is about 350 °C. Transition from the liquid to the plastic state takes place between 320 and 280 °C, whereas the extrusion of the lead is performed at a temperature between about 200 and 180 °C.

The lead is supplied by the metallurgical works in the form of pigs of about 99.8 per cent purity; from the melting kettle it is fed directly to the container. Scrap lead is often added to the lead bath in order to reduce the price of the pipes. Pure lead gives aboud 2% lead dross, whereas about 8% residue have to be taken into account in scrap lead. The dross is collected and refined by chemical works. Larger quantities of scrap lead are refined in a simple manner by heating it to a temperature of about 450 to 500 °C and skimming off the dross from the surface of the metal.

For a long time there was an ever increasing demand for lead pipe presses. However, this development came to a stop about fifteen to twenty years ago, when pipes made of steel, non-ferrous and light metals, as well as pipes of plastic or synthetic materials made their appearance on the market.

Apart from the lead pipe press, there was also designed a series of smaller presses, the most important types of which being employed for the manufacture of curved lead pipes for use as bends or syphons, of lead wire of round or profiled cross-section, and of tin solder wire.

a) Lead Pipe Presses

The first patent on a lead pipe press was granted to the Englishman JOSEPH BRAMAH as early as 1797. The design of this press is shown in Fig. 1[1]). Liquid lead flows from a melting kettle a, which is heated from beneath, through the valve c into the cylinder when the piston b moves up. Piston b moving down causes the lead to be forced trough an annulus

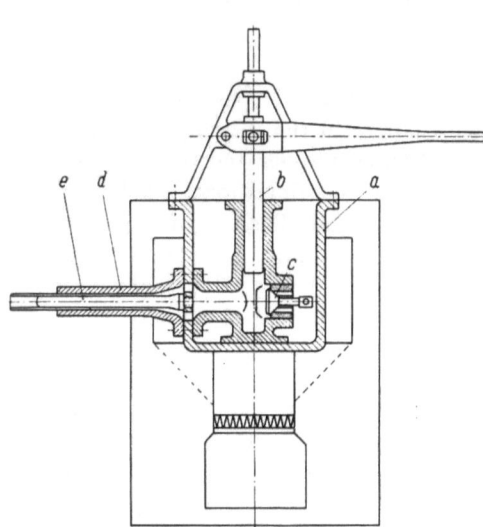

formed by the tube-shaped projection d and the mandrel e. The mandrel is secured to the kettle casing which is provided with several slots to allow for the passage of the lead to the tube-projection. At the extreme end of the annulus the lead solidifies and issues in the form of a pipe. Fuel gases are circulated around part of the tube projection d.

Decades had still to pass since the publication of this invention, until the first hydraulic presses were constructed and it was not until 1867 that the lead

Fig. 1. BRAMAH's lead pipe press

pipe press, as shown in Fig. 2, by the Frenchman HAMON became known. The principles of this early invention are still embodied in presses at the present time.

In order to render possible casting of the molten lead into the container, a lifting cylinder a with piston b for the lifting of the crosshead c is arranged on top of the press. The nuts d are provided with bayonet-type threads. When changing tools, the crosshead f may be lifted with the help of the clamps e. Casting of the liquid lead is followed by an interval of five to ten minutes to allow for the solidification of the metal. The extrusion temperature is about 210 °C.

Fig. 3 shows the design of a modern lead pipe press. It has a bottom cylinder platen a which is positively connected to the top counterplaten c by means of four columns b. The plunger d runs in a bronze bush in the cylinder and travels over a mandrel-bar holder e which is rigidly secured in the base of the cylinder. Sealing of the plunger in the cylinder and on

[1]) PEARSON, C. E.: The Extrusion of Metals. London: Chapman & Hall 1953.

the mandrel-bar holder is ensured by leather- or vulcanized fiber packing collars which may be removed after detaching a stuffing box flange and a screwed union. The mandrel-bar f is screwed to the top end of the mandrel-bar holder.

The plunger-head is guided in its travel on the columns by a crosshead g and carries the hollow extrusion ram h. The latter is secured to the crosshead by means of a flange and may be centered by means of four lateral screws. A dummy block of forged high-tensile bronze rests on the ram. Two drawback cylinders i with stationary drawback plungers j being attached to the counterplaten, are arranged in the plunger crosshead to obtain fast lowering speeds of the ram. The container k is inserted in the counterplaten and secured by a flange. Heat-insulated ducts, to which is connected a gas burner, are provided in the counterplaten for the heating of the container. Heating may also be effected electrically.

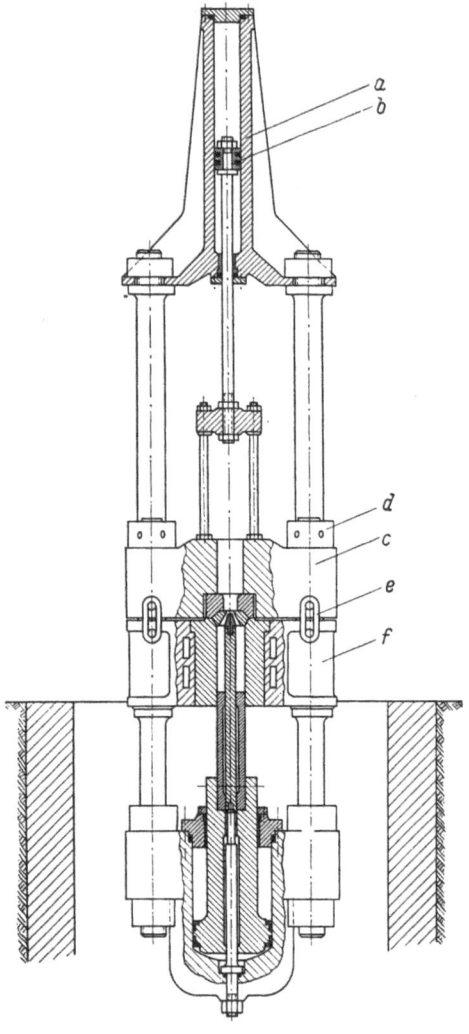

The tooling consists of the mandrel-tip, die, die-ring and die-holder, the latter being locked in the platen by two opposed wedges l. The wedges are adjusted via a spindle m with left- and right-hand thread by turning a

Fig. 2. Lead pipe press by HAMON

handwheel n. The emerging lead pipe runs over a pulley o to a coiling drum. Pipes which cannot be coiled on account of their large diameter are extruded vertically towards the head and cut off directly above the counterplaten. The vertically extruded pipe is lifted by a crane (winch) as the following hot pipe end cannot support the extruded pipe over the die.

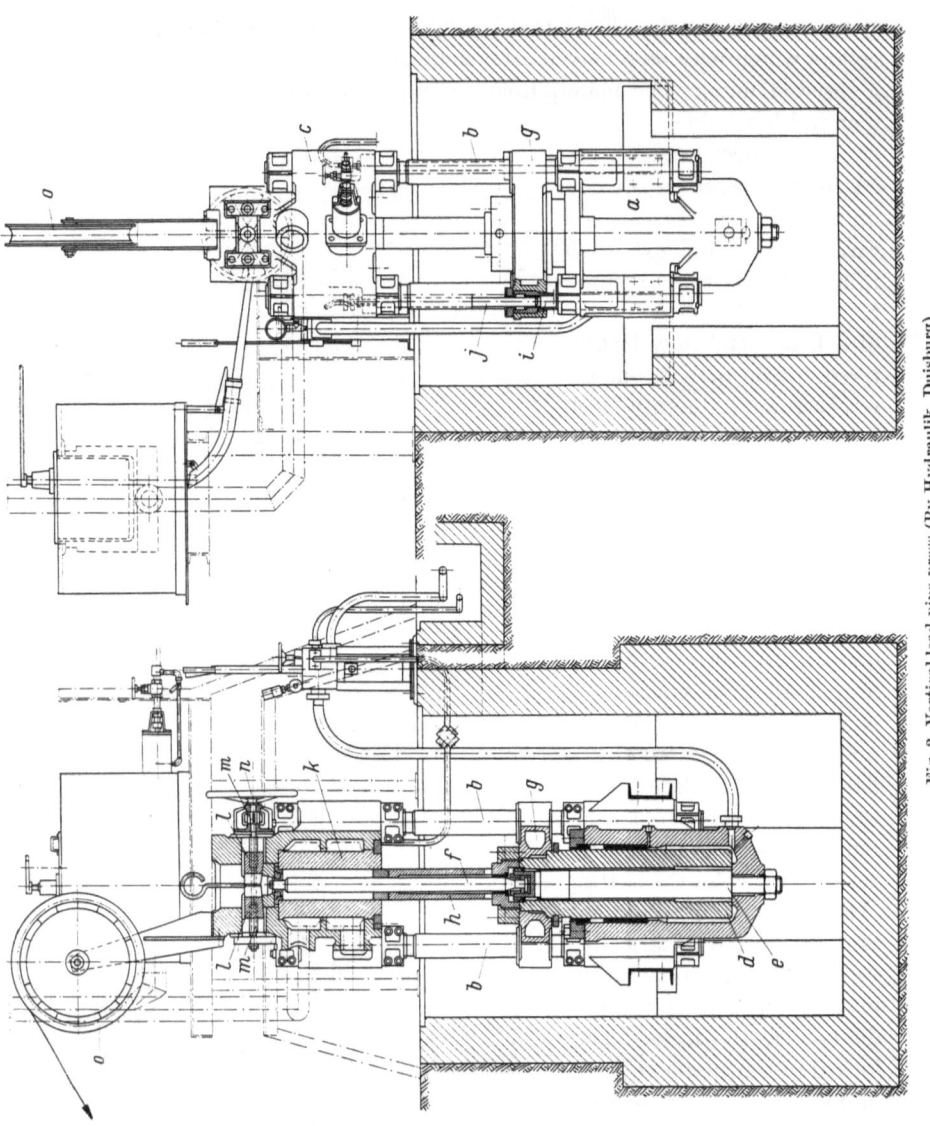

Fig. 3. Vertical lead pipe press (By Hydraulik, Duisburg)

The commercial weight of a pipe bundle is about 75 kg with a maximum pipe diameter of about 60 mm. The bundle is lifted out of the coiler, wrapped with straw ropes and prepared for shipment. For the purpose of roughly determining the weight of a pipe bundle, the press is provided with a dial gauge indicating the extruded weight in dependence of the ram stroke.

At the end of the operation the die-holder is unlocked and die-holder with die and residual slug are ejected by lifting the ram. The die-holder is provided with two eyebolts into which a round rod is inserted for the purpose of lifting the former. In order to make the tooling easily accessible and to facilitate the operation of the press, the latter is placed in a pit or cellar so that the top edge of the counterplaten is about 800 mm above floor level. The lead is molten in a melting-pot, which is placed at the side of the press, and flows through a trough into the container. The melting-pot is generally placed on a platform and a poppet-valve is mounted in its base. Heating is done by gas or electricity.

Charging the press with liquid lead is disadvantageous in that there must be allowed for a setting period between the cycles. Extrusion of preheated hollow billets would obviate this setting period, thus achieving a considerable increase in press capacity. This, however, meets with difficulties. It has been found that there are often heavy inclusions of oxides in the junctions between the billets which cause defects in the extruded pipes. Apart from this it is not easy to evacuate the air from the container. If an air-bubble is formed in the lead, it travels through the die under a very high compression and on leaving the die it suddenly expands, thus frequently causing a hole in the wall of the pipe.

Power and stroke of a lead pipe press generally depend on the lead charge, this being determined by length, diameter, and wall thickness of the largest pipe to be extruded, to which value there are added 15 to 20% to cover for slug, discard, etc. The length of the pipe in turn depends on the required weight of the pipe bundle. The mandrel-bar is to be about 10 mm wider in diameter than the maximum inside diameter of the pipe. The actual pressure capacity is therefore given by the equation

$$P = \left(\frac{\pi D^2}{4} - \frac{\pi d^2}{4} \right) p$$

in which D is the diameter of the container, d the diameter of the mandrel bar, and p the extrusion pressure, including all flow-, frictional-, and hydraulic resistances.

An extrusion pressure $p = 1,600 \text{ kg/cm}^2$ is chosen to allow for alloys; for pure lead an extrusion pressure $p = 1,200 \text{ kg/cm}^2$ would be sufficient.

The container diameter D is given by the equation of the weight G of the lead charge

$$G = \left(\frac{\pi D^2}{4} - \frac{\pi d^2}{4} \right) h \gamma$$

in which γ is the specific gravity of the lead charge and h the effective stroke which is assumed as $h = 3$ to $4 D$.

When extruding very large diameter pipes in straight lengths, it is general practice to use another thicker mandrel bar. Due to the cross-section of the annulus between bar and container wall being decreased, the actual power of the press is in such case not fully utilized. The resulting reduction in the weight of the lead charge is no drawback, as in general the charge is still large enough to allow for the extrusion of several straight lengths of pipe. This technique is, however, recommended only when there is an occasional demand for small quantities of such large pipes.

In size lead pipe presses range from 250 to 1,250 tons total pressure capacity. Materials employed are cast steel GS 50 for cylinder-platens, GS 45 for counterplatens and crossheads, steel St 60 for plungers and

Table 1. Vertical Lead Pipe Presses

Pressure capacity, tons	250		450		630		1,250	
Pipe diameter, mm	5 to 70	60 to 100	10 to 90	80 to 150	25 to 140	130 to 200	30 to 150	150 to 300
Container diameter, mm	160	160	200	200	265	265	340	440
Mandrel bar diameter, mm	80	115	100	152	155	205	165	320
Lead charge, kg	70	50	160	90	420	250	1,000	1,000
Cycles per hour	5		4		2.5		2	
Max. ram speed, mm/sec	4		4.5		5		5	
Pump capacity, H. p.	20		40		65		130	

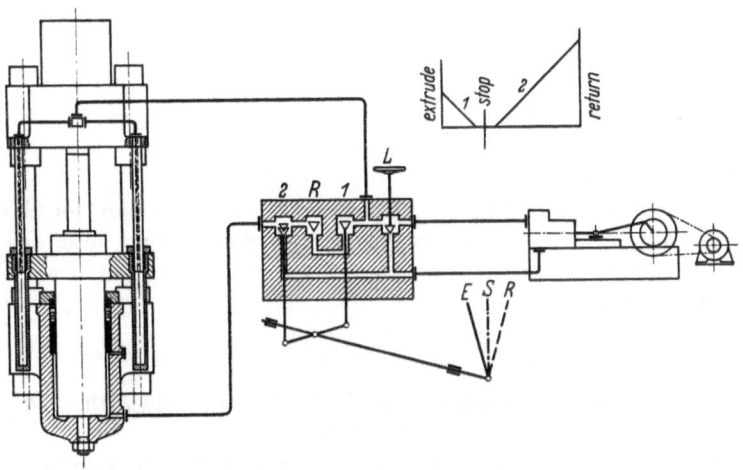

Fig. 4. Hydraulic circuit diagram of a lead pipe press with direct pump drive

plunger rods, St 50 for columns, St 70 for containers, rams and mandrel rods, and alloy steels for mandrel-tips and dies.

Table 1 shows principal dimensions of presses built.

Fig. 4 illustrates the hydraulic circuit diagram of a lead pipe press driven directly by pumps. The control gear is provided with control valves *1* and *2* for the inlet and outlet of the pressure water, a check valve *R* arranged between the two former valves, and a regulating and by-pass valve *L*, mounted in front of the inlet valve. When valve *L* is open, the pump idles. The press maintains its stroke position during idling, as the check valve *R* prevents a descent of the main plunger. Partial opening of valve *L* causes the water to be throttled, thus permitting of regulating the extrusion speed at random. The pump output is dependent on the ram speed. Time-tried performance figures are also shown in Table 1. Instead of operation by pressure water, the presses may also be arranged for pressure oil operation.

b) Lead Trap Presses

The use of lead pipes for drains soon led to the need of curved pipes – so-called syphons or traps – consisting of a U-shaped bend with sub-

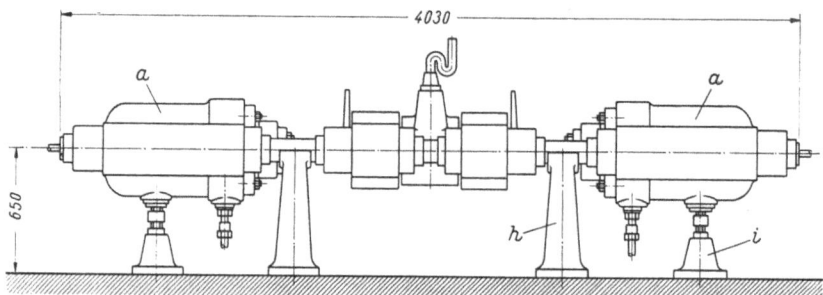

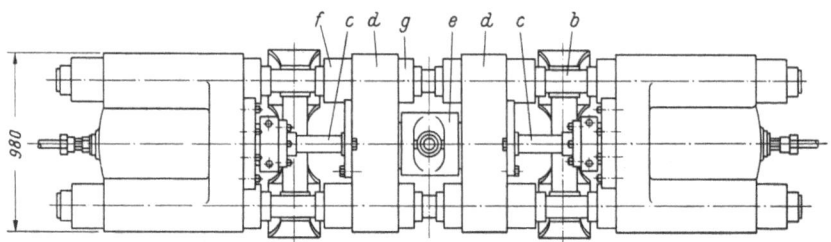

Fig. 5. Lead trap press (By Breda Fucine, Milan)

sequent elbow and serving to close the conduit by a water-trap formed in the bend, thus preventing any vapors or gases to flow back causing

unpleasant odors. At the present time these syphons are made of synthetic compounds, too, whereby the application of these presses has been
widely diminished. The operation of such presses is of particular interest,
because it shows how a curved pipe can be extruded by simply controlling the flow of lead.

The design of a lead trap press is illustrated in Fig. 5. Contrary to
lead pipe presses, they are in horizontal design. They are provided with
two main cylinders *a* being connected to each other by means of two columns *b*. The rams *c* are attached to the double-acting pistons. The two crossheads *d* accomodate the containers with the die-block being arranged between the former two. The crossheads are braced by nuts *f* and jam nuts *g* in such a way as to permit of applying the pressure unila-terally, too. The press rests on two supports *h* and the spindle supports *i*.

In the operation of lead trap presses, as shown in Fig. 6, molten lead is supplied through the orifices *k* to the horizontal containers, these being closed by the rams. When the rams – after solidification of the metal – advance at a uniform rate of speed, the lead is forc-ed, inside the die-block *e*, through two semicircular pas-sages *l* shaped and finished by hand, and through four bores *m*, into the extrusion chamber

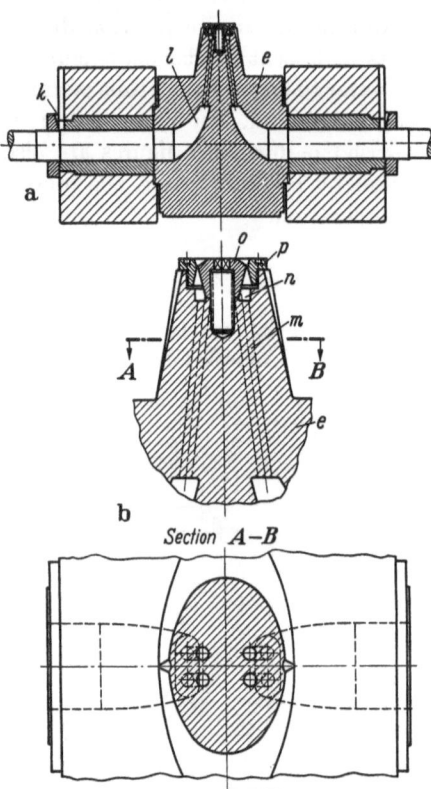

Fig. 6a and b. Tooling of the lead trap press shown in
Fig. 5

n, and over the stationary mandrel *o* through the die *p* to form a
straight pipe. By advancing one ram faster than the other, a curved
pipe is produced, the curvature increasing with the difference between
the two ram speeds being increased.

Trap presses are operated like pipe presses directly off one pump
which – in connection with a control gear – delivers the pressure water
or the pressure oil either simultaneously to the two main cylinders or

alternately to one or the other. The control gear must permit a high-precision regulation of the delivery volume and its operation calls for high skill on the part of the operator.

Time-tried lead trap presses work at 300 tons pressure capacity and a lead charge of 30 kg with the container diameter being 82 mm, the stroke 2 × 250 mm and the pump capacity being about 22 H.P.

One-cylinder presses in vertical design, in which the flow of the lead through the die is regulated by displacing a diaphragm laterally, have given less satisfactory results.

c) Lead Wire Presses

Lead wire may be produced in presses of relatively simple design. It is remarkable that the indirect or inverted method is in most cases applied when extruding wire. With this method the lead is not extruded through a die mounted in the base of the container, but the die is fixed on the end of the ram. The ram is bored out to allow for the wire to issue through the top of the ram, i. e. the wire moves in the opposite direction of the ram movement. Thus, the extrusion resistance is reduced considerably and the charge may therefore be much larger than that employed in the direct method of extrusion (see also p. 136). For the purpose of increasing the production of the press, it is advantageous to work with two interchangeable containers, whereby the setting period of the liquid lead is gained.

The press is suitably of the down-stroking two-column type, as shown in Fig. 7. The two columns are connected to the cylinder- and bottom platens by split nuts and check nuts. The pullback cylinder is suspended in the main cylinder so as to reduce the overall height. The pullback plunger acts on a crosshead which is connected to the plunger cross-head by two tie-rods. The plunger head carries the extrusion ram, on the bottom of which is fixed the die. The ram is bored out and provided at its bottom end with a curved outlet, through which it is possible to push out the die with the help of a rod.

The two containers are mounted in a rotating arm which is rotated about one of the columns and rests on a ball-bearing.

In the extrusion position the claws on the bottom side of the rotating arm slip under two collars on the bottom platen, thus preventing the container from raising when the extrusion ram is retracted. The undersides of the two containers slide over a table consisting of a ring-shaped sheet-metal plate and being supported at the charging side by a bracket fixed to the bottom platen and for the remaining part by structural sections on the foundation. When the extrusion ram is retracted, the container is held in the bore of the rotating arm by a shoulder. The arm is

rotated by hand with the help of a rod. The containers are insulated with
glass wool to prevent radiation of heat.

Lead wire presses are constructed for pressure capacities of about
250 tons, while the ram diameter is 90 mm and the effective stroke

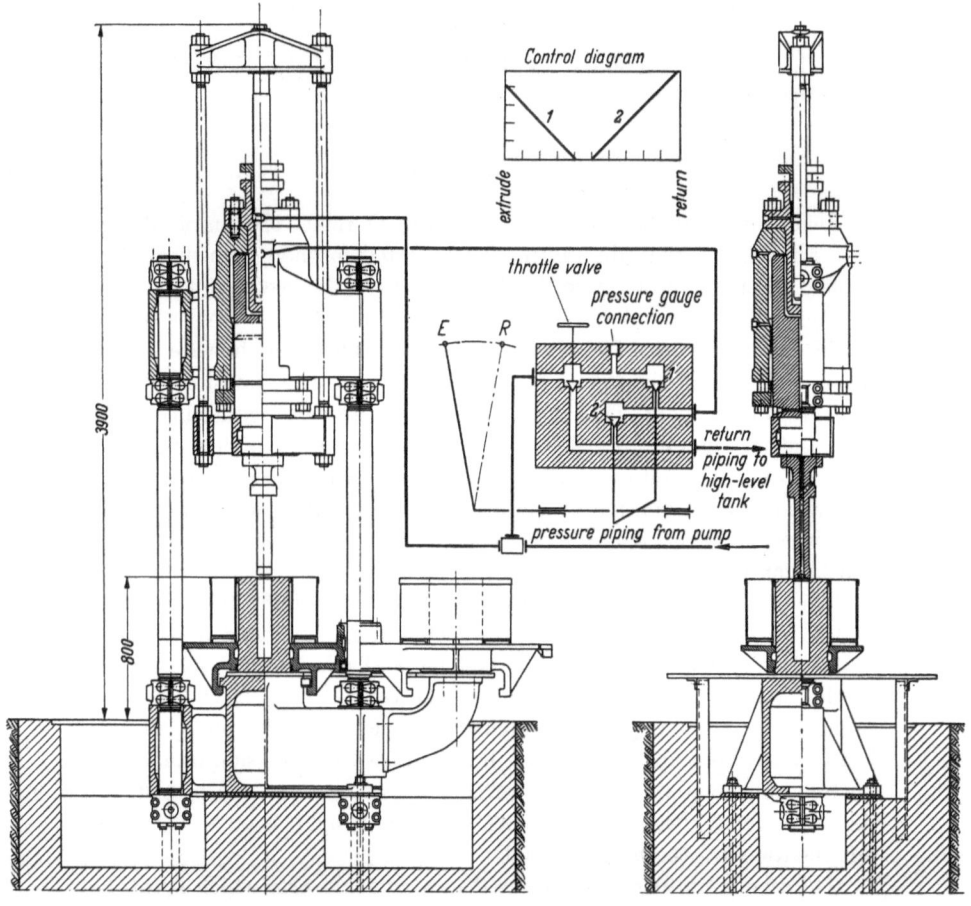

Fig. 7. Vertical lead wire press with rotary arm accomodating two containers

amounts to 450 mm, The effective charges are 30 kg and the ram pressure
on the lead is 3,800 kg/cm². This pressure – as compared with the values
given for lead pipe presses on page 5 – is relatively high, thus allowing,
however, for the extrusion of lead alloys.

Like lead pipe presses, the lead wire presses are driven directly off
the pumps. The pump output is dependent on the ram speed and the

delivery speed of the wire respectively, which is assumed at 60–120 m/min; the pump capacity is about 20 H.P.

Fig. 7 furthermore illustrates the operation of the control gear. The delivery line of the pump leads both to the control gear and directly to the pullback cylinder. The valve housing accomodates an inlet- and an outlet valve 1 and 2, as well as a throttle valve which is adjusted by means of a handwheel through a spindle. Valves 1 and 2 are operated alternately by a hand-lever through a cam shaft, as indicated in the valve-lift diagram. When valve 1 is closed, the return movement takes place, which may be regulated at random over its full stroke length by means of the throttle valve. Valve 1 being opened and valve 2 being closed, the extrusion cycle starts. It is also possible to vary the working- or wire speed by means of the throttle valve. If a high resistance is encountered at the end of the stroke, the pump is automatically set to idling by a valve releasing device[1]) as soon as the final pressure is exceeded.

d) Solder Wire Presses

Solder wire consists of a tin tube of a maximum outside diameter of about 9 mm and a wall thickness of 1.5 to 2 mm, which is filled, while being extruded, with hot, liquid colophony. The wire is extruded on vertical or horizontal presses.

A vertical type of press is shown in Fig. 8. The upper part of the steel-plate frame accomodates the main cylinder with its plunger, the latter being sealed by packing collars and an adjustable stuffing box. The plunger forces down the crosshead, being guided in its travel on the frame by adjustable gibs. Two return cylinders are arranged in the bottom part of the frame, the plungers of which press against the crosshead from below.

The container, consisting of a liner with shrunk-on jacket, is placed on the table plate. The liquid tin is supplied to the container through a hopper and impurities are skimmed off the surface after filling. The extrusion ram, loaded by the weight of the moving parts, rests on the charge while it solidifies, thus preventing the formation of a contraction pipe. During the extrusion a thin shell is formed between ram and container which corresponds to the existing clearance in diameter of 0.3 to 0.5 mm. The ram may be exactly aligned to the center-line of the container by means of setscrews on the ram flange.

The design of the tools in the base of the container is illustrated in Fig. 9. The tin is extruded vertically to the center-line of the container,

---

[1]) MÜLLER, E.: Hydraulische Pressen Bd. 2. Berlin/Göttingen/Heidelberg: Springer 1955.

hence horizontally, by forcing it through the annulus formed by die a and the mandrel b. The mandrel is provided with a bore c, the diameter of which is about 1.5 mm at the tip of the mandrel. The bore expands

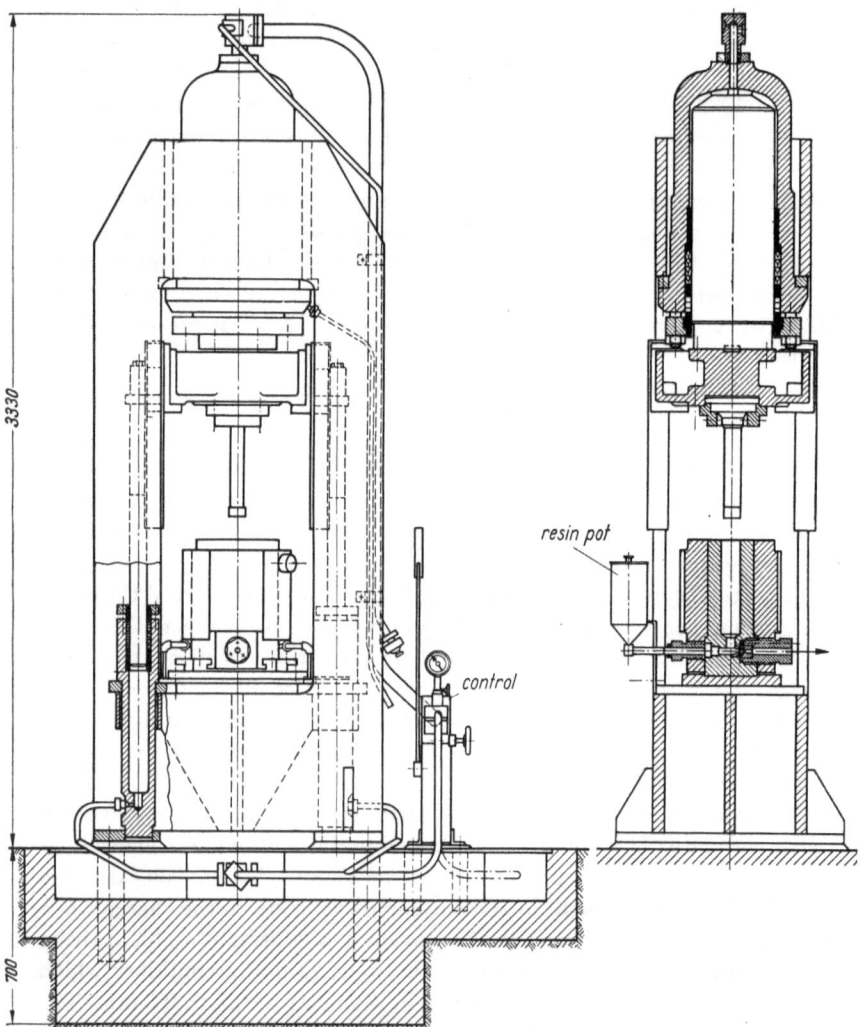

Fig. 8. Vertical press for manufacture of solder wire with resin core (By Hydraulik, Duisburg)

to the rear and serves as inlet for the colophony. The die is set in the die-holder d which can be moved into any direction by three wedge bolts e, thus permitting of adjusting the wall thickness of the tin tube accurately.

A steam- or electrically heated pot contains the colophony which is supplied to the mandrel through a pipe. Container and tooling are heated, too, and for this purpose the jacket is provided with a number of bores

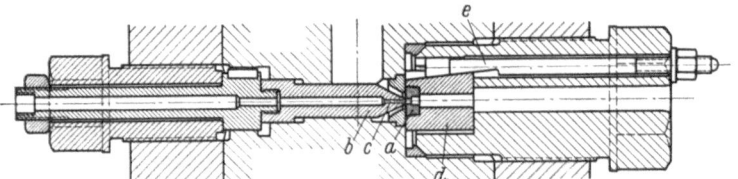

Fig. 9. Tooling arrangement for press shown in Fig. 8

of about 20 mm diameter each, to accomodate the heating coils. The temperature of the container assembly is suitably controlled by means of thermocouples; during extrusion it must be about 95 to 110 °C.

Fig. 10. 250-ton horizontal press for manufacture of solder wire with resin core
(By Schloemann, Düsseldorf)

Fig. 10 shows a solder wire press of the horizontal type. The main cylinder is connected to the counterplaten by two columns. The plunger, which is sealed by a stuffing box and packing collars, presses on a crosshead guided in its travel on the columns, and is retracted by a pullback piston arranged behind the cylinder. The extrusion ram is attached to the crosshead; it is exactly aligned to the center-line of the container by means of centering screws. The container is inserted in the counter-platen and insulated to prevent radiation of heat.

The press is set on a base frame in a position inclined from the horizontal so as to permit the escape of air while the liquid metal is charged and to allow for impurities to rise to the surface. The colophony pot is mounted on the counterplaten.

The general arrangement of the tools with the container and the ram for extruding solid wire is illustrated in Fig. 11. While filling the con-

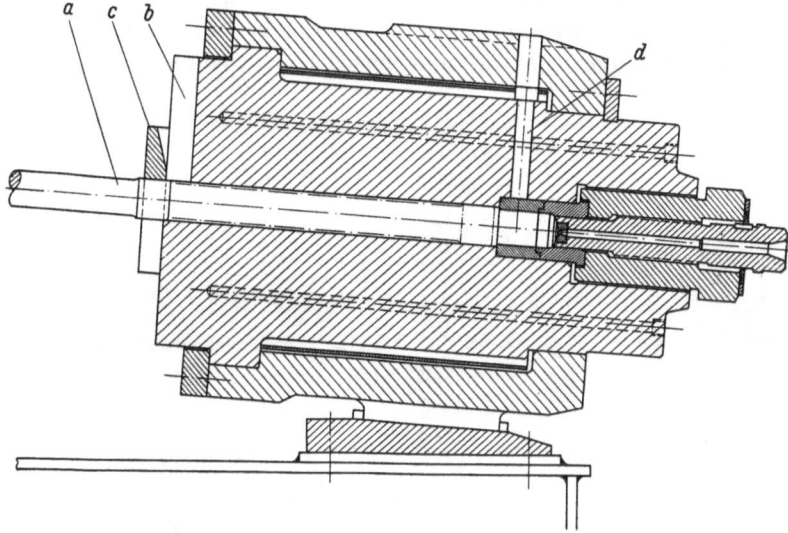

Fig. 11. Tooling used for the extrusion of solid wire

tainer, its front end is closed by the ram a. With the help of a funnel the metal is passed through slot b of the container. The slot is covered at its front end by a horseshoe-shaped swing diaphragm d which is arranged around the ram head. The swing diaphragm serves also to limit the stroke of the ram-holder on completion of the cycle. In case the tools are to be removed, the swing diaphragm is removed by traversing it by means of a rod. Now the ram can perform an excess stroke and, the rear screwed joints having been detached, easily eject the tool assembly.

For the manufacture of colophony core wire, the colophony feed pipe is inserted in bore d and a mandrel with mandrel-holder, similar to that shown in Fig. 12, is employed. With the arrangement being as shown in Fig. 12, the mandrel-holder a is inserted in the container bore from the rear; it is provided with channels b to allow for the passage of the tin, as well as a bore c for the inlet of the colophony. The mandrel d is exchangeable and screwed in. The die e is set in a die-holder f which is axially adjustable to obtain different wall thicknesses. The sleeve g is provided with a nose h for the stripping of the discard when detaching the screwed joint i. Container and colophony pot are heated electrically. The container is provided with long bores to accomodate the heating cartridges. The operating temperature of the container is about 100 °C.

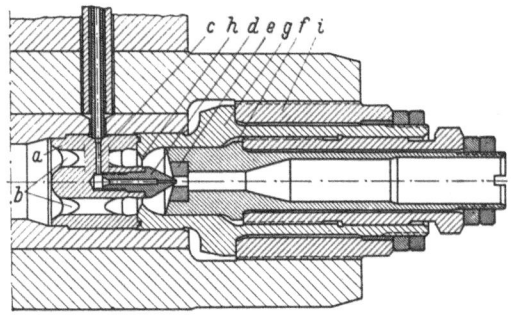

Fig. 12. Tooling used for the extrusion of solder wire with resin core

Solder wire presses are generally designed for a pressure capacity of about 250 tons and an effective charge of about 10 kg. The maximum ram pressure is about 6,500 kg/cm² and ratio $d : l = 1 : 5$, in which d = diameter of container and l = effective length of container. The maximum diameter of the dies for the extrusion of tubes is about 20 mm.

The press is driven – like wire- and pipe presses – directly by a pressure water or pressure oil pump at a maximum operating pressure of about 315 atm. The pump output is about 10 to 12 H.P. giving a ram speed of about 1.4 mm/sec; the control gear is as described on p. 10.

Chapter II

PRESSES FOR THE MANUFACTURE OF LEAD AND ALUMINUM CABLE SHEATHING

Amongst lead working presses, the most important position is held – both in number and value – by the cable sheathing presses. Their design – in particular that of their tooling – has gone through a long development, the history of which is, however, outside the scope of this book and it is therefore not intended to depict it here in detail.

In Germany the first lead cables were provided by WERNER SIEMENS[1] as long ago as 1851 for the fire-alarm network of the City of

---

[1] BORCHARDT, H., u. S. v. WEIHER: 75 Jahre Kabelwerk Berlin. Siemens-Schuckertwerke AG 1951.

Berlin. These cables consisted of a gutta-percha covered conductor thread-ed into a lead pipe, the diameter of which was subsequently reduced by cold drawing until the pipe fitted tightly around the cable core. Three decades had to pass until in 1882 there was used a press, constructed by WERNER VON SIEMENS, on which the lead sheath was directly extruded on to the cable core.

The press used was of the vertical type (see p. 3) and the cable core was passed up from below through a bore in the mandrel-bar and the mandrel into the die where the sheath was formed which then gripped and entrained the cable core. The drawback of this method was, how-ever, that only relatively short lengths could be extruded in one cycle, because the press was charged with cold, hollow-cast billets. Considerable improvement was achieved when the method of charging liquid lead and extruding charges in succession was introduced. Thus endless lengths of cable could be manufactured and this method has been maintained up to the present time.

Despite its many drawbacks the lead sheath has found widest ap-plication in the course of the decades up to the present time to protect the cable core against the entry of liquids and vapors and – frequently in connection with a suitable reinforcing – against mechanical stresses. In some special cases the sheath served as grounding wire or neutral conductor, too.

Its drawbacks are as follows:

1. The lead used for the sheath represents a high percentage of the total weight of the cable due to the high specific gravity of the lead; it runs at an average of 30 to 50% and, in extreme cases, it may be as high as 80% of the cable.

2. In spite of the addition of alloying components – such as tin, anti-mony, tellurium, etc. -- the mechanical strength of the sheath is low. Thus it is highly liable to damage and furthermore it is sensitive to inter-nal pressures imposed in oil-filled high-voltage cables.

3. Its tendency to recrystallize even in ambient temperature, this being furthered by such factors as heat through Joule effect, exposure to sun-light, etc.; thus a crystalline growth occurs which in the event of vibration stresses, set up for example when laying a cable along a bridge, very easily leads to the formation of splits, thus rendering the cable unserviceable.

It is on this account that in the course of time numerous experiments were carried out with other materials with a view to improving the pro-perties of the sheathing and at the present time sheathing is even made of synthetic materials, steel and aluminum. However, in the manufacture of such sheathing, the use of hydraulic presses is limited to the extrusion of aluminum sheathing only.

Manufacture of aluminum sheathing[1]) was commenced about 25 years ago. To begin with, the most successful method was that employed for the early lead sheathed cables, i. e. the core was threaded into an aluminum pipe which was subsequently drawn through a die to reduce its diameter. Disadvantages experienced with this method are that cold working embrittles the pipe and, apart from this, a working space of several hundred meters length is required, as the pipe has to be wound off a drum prior to threading the cable core. Very soon experiments were made to apply the direct method of extrusion as used in the manufacture of lead sheathing. Difficulties arising with this method, were mainly due to the high temperature of about 450 °C, which is required for the extrusion of aluminum. This easily leads to a charring of the cable insulation. Most of these troubles have been eliminated now and presses for the extrusion of aluminum sheathing (see Fig. 13) have found a wide application.

Since the construction of the first lead cable press it has been one of the main aims to invent a continuously working press which leaves no stop marks on the cable sheath. Stop marks are produced when the press is restarted after completion of a cycle and solidification of the fresh charge, when a ring, the so-called bamboo-ring, appears on that part of the sheath which is just emerging from the die. At this point the cable sheath is invariably weaker – though to a minor degree only – this being caused by a variation in the wall thickness or by a notch effect.

The simplest means of obviating stop marks would be to employ a container holding such a quantity of lead as would be required for the manufacture of the largest and longest cable sheath. This would mean, however, the use of a very powerful and big press, the price of which would be extremely high so that for this very reason cable manufacturers rather prefer to work with a smaller press and to put up with the stop marks.

Many suggestions for the design of a continuous lead cable press have been published. First success was achieved after many experiments on screw presses which were already known in the cable industry for the manufacture of sheathing of synthetic materials. Their design having been considerably changed and being considerably heavier, these presses are also suitable for charging with liquid lead. However, experiments to run them on alloys containing high additions of antimony, tellurium etc., have been unsuccessful up to the present time.

---

[1]) HANFF, F., G. HOSSE u. W. DEISINGER: Aluminium als Baustoff für Kabelmäntel, Siemens-Zeitschrift Bd. 19 (1939) H. 8. – BORCHARDT, H., u. G. HOSSE: Kabel mit gepreßten Aluminiummänteln. Siemens-Zeitschrift Bd. 27 (1953) H. 4.

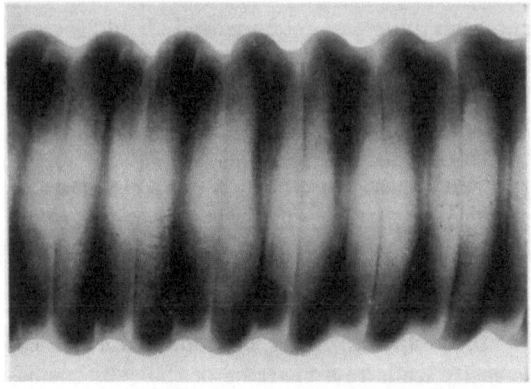

Fig. 13a. Corrugated aluminum sheath with stop mark
in center of picture after 18
bends over the 20-fold cable
diameter. Outside diameter
of sheath 97 mm

Fig. 13b. 6-kV power cable
of 3 × 95 mm² with smooth
aluminum sheath

Fig. 13c. 10-kV power cable
of 3 × 70 mm² with smooth
aluminum sheath

Fig. 13d. 1-kV power cable
of 3 × 300 mm² with corrugated aluminum sheath

Fig. 13e.   Three-conductor
oil-filled cable for 110 kV,
3 × 150 mm², outside dia-
meter of cable about 95 mm

Fig. 13a–e. Aluminum Sheathed Cables (Courtesy of Felten & Guilleaume Carlswerk AG,
Köln-Mülheim)

Recently there has been developed a continuous hydraulic ram press which, however, is used in the first place for the extrusion of aluminum sheathing, since the stop marks in aluminum sheathing are much more sensitive than those in lead sheathing.

a) Lead Cable Presses for Charging of Liquid Metal

One of the first cable presses which was widely adopted both in Germany and in foreign countries, is the Huber Press, brought out in 1892 (Fig. 14) and named after its inventor, which continues to be in operation

Fig. 14. 575-ton horizontal two-ram lead cable press, Huber system
(By Krupp Grusonwerk, Magdeburg)

even at the present time in a number of works. This is the first example of arranging the axis of the press at right angles to that of the cable passage in order to simplify and facilitate access to the tooling. This arrangement has given very satisfactory results and has been maintained in principle in all presses ultimately developed.

The tooling is illustrated in Fig. 15 and consists of a die-block a and two rams b acting opposite to each other. The rams force the lead through an annulus formed by the mandrel c and the die d. The lead pipe e thus being formed fits round the cable core, which is indicated by a dash-dot line, and draws it through the hollow mandrel. The sheathed cable core is coiled on a drum, placed at a distance of a few meters from the delivery point. Mandrel and die must be exchanged to suit the respective cable diameters. In order to keep down material expenditure, the mandrel is inserted in a mandrel-holder f. The back-pressure, which occurs during extrusion, is taken up by the screwed joint g of the mandrel, this being secured in the die-block by means of a buttress-type thread so as to be easily detachable. The wall thickness of the cable sheath is regulated by tightening or releasing the mandrel screwed joint.

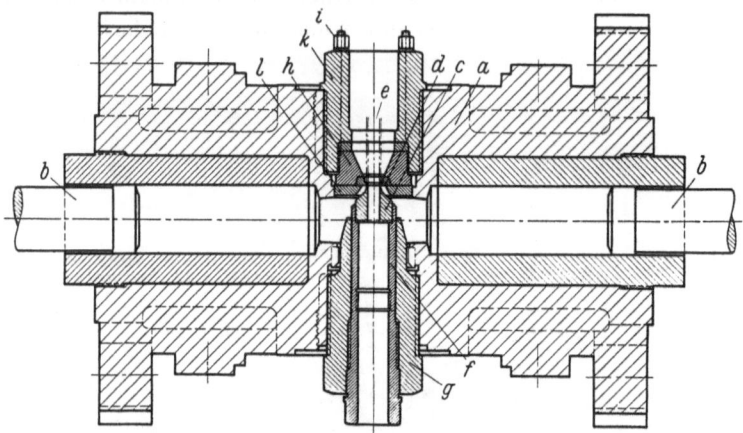

Fig. 15. Tooling for a horizontal two-ram lead cable press

As the mandrel is held in the mandrel-holder, the die is set in the die-holder h. The die-holder is radially adjustable by three wedge bolts i to afford a means of controlling the concentricity of the sheath. The back-pressure exerted on the die-holder, is taken up by a screwed joint k which is always rigidly tightened. In front of the die there is arranged a base ring l serving to regulate the flow of lead. A slight accumulation of lead takes place in the base ring, thus making up for minor sources of error caused by the rams not moving at a uniform rate of speed and by inaccurate shaping and finishing of the die-block.

Fig. 14 shows the tooling in connection with the press. The latter is of horizontal design, having one main cylinder on the left-hand side and another one on the right-hand side, these two cylinders being positively connected to each other by four columns. In either cylinder moves a double-acting piston, the piston rods of which being sealed by a stuffing box and carrying the rams. The die-block is enclosed by a jacket, which at its bottom is connected to a heating furnace and on its top carries the lead melting kettle. At that time the furnace was coke-, gas- or oil-fired. The control gear is arranged at the right-hand side of the press. Turning of a handwheel operates a diaphragm whereby the press is controlled to advance or return. Another diaphragm operates as a function of the piston movements and when it moves forward, it throttles the respective cylinder side, thus ensuring a relatively accurate and uniform opposed movement.

The press rests on its full length on a base frame. The tools are extraordinarily easily accessible and the cable enters at and emerges from the center of the press vertically to its longitudinal axis. A worm-driven rotating mechanism serves to release the screwed joints for mandrel and die when changing the tooling.

The press was built in two sizes, the data of which are given in Table 2.

Table 2. Horizontal Lead Cable Presses (Huber System)
(Pure Lead and Alloy Lead with up to 3% Tin)

Pressure Capacity tons	Operating Water Pressure atm	Charge kg	Max. Cable Dia. mm	Lead Displacement kg/min	Pump Capacity	
					l/min	H.p.
350	300	100	65	22	20	18
575	300	200	120	41	40	36

As time passed on there was an ever increasing demand for cable presses capable of holding larger charges. During this development considerable difficulties were met with the horizontal design, which were mainly caused by the two driving units and by the unfavorable arrangement of the containers complicating filling of the liquid lead. Thus, the idea of a vertical design was evolved. On account of the necessity to protect the cable core against leak water and with due regard to the low headroom available in most of the plants, preference was first given to the up-acting press type, i. e. a press with hydraulic operating cylinders arranged in its base, with moving die-block and container, and with fixed extrusion ram. An early example of this type is illustrated in Fig. 16.

The press takes the form of a cast-steel frame, in the base of which is inserted the hydraulic cylinder. The main piston is a simple plunger piston, moving in the cylinder in a long bronze bush and sealed by a stuffing box. The plunger carries the die-block with the container arranged on top of it. The latter is set in a crosshead, to the sides of which two return plungers engage. The return force acts in the same sense as the weight of all of the moving parts, which on its own, however, would not be sufficient to pull the container off the ram. The crosshead is reliably guided in its travel on prismatic guides on the frame. It is connected to the plunger head plate by solid bolts. The extrusion ram is fixed and secured to the upper chord of the frame by a flange, in which a number of setscrews is provided to permit of centering the ram in the container bore.

Two-column presses, as shown in Fig. 17, have found as wide an application as presses of the frame type. Preference is given to the former, if

maximum charges and maximum operating pressures are required. A closed frame is formed by the two columns, the bottom hydraulic cylinder and the top crosshead. It is advantageous to choose a press design in which the column ends are under a permanent prestress. For this purpose the check nuts on the cylinder-platen and on the crosshead are screwed down – after assembly of the press – while the pressure in the main cylinder is increased about 10%.

Fig. 16. 600-ton vertical lead cable press in frame construction, with underlying hydraulic cylinder, moving container and stationary extrusion ram arranged in the head of the frame (By Krupp Grusonwerk, Magdeburg)

The columns also serve to guide the container crosshead or the return crosshead respectively, in its travel. In some cases – especially if heavy presses were concerned – four-column presses have been built, which afford better stability. An example of this type is shown in Fig. 18. Stability being of minor importance only on account of the slow operation and the central load in cable presses, preference is in general given to two-column designs, because they afford easier access to the tooling.

While in the horizontal two-ram press the two streams of lead come together in the die-block in the central plane of the mandrel, where they weld and form half of the sheath each, in the single-ram press the lead must be forced round the full circumference of the mandrel, thus rendering the lead flow more difficult and increasing the power input by about 20 to 25%. Inspite of these disadvantages, the advantages obtained in

Fig. 17. 1,400-ton vertical two-column lead cable press with underlying hydraulic cylinder and stationary extrusion ram arranged in the head part (By Hydraulik, Duisburg)

Fig. 18. 2,000-ton vertical lead cable press, as per Fig. 17; however, in four-column design (By Hydraulik, Duisburg)

the vertical design are so obvious that the horizontal design has been completely abandoned for presses to be charged with liquid metal.

Fig. 19 and 20 illustrate the die-block of the early vertical presses. It consists of a bottom part a and a top part b, these being centered and connected to each other by four bolts. Splitting of the die-block is necessary for the purpose of shaping and finishing the passage from the

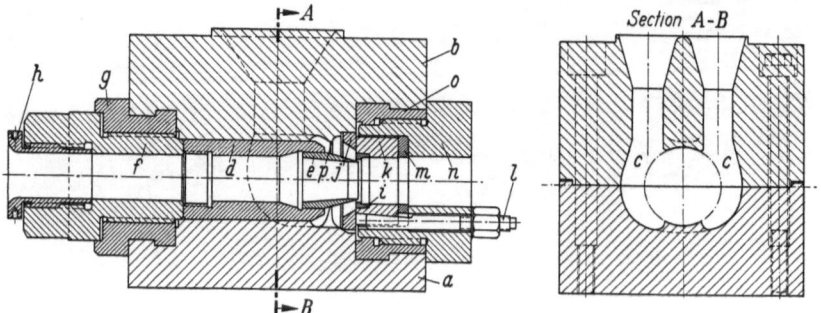

Fig. 19. Split two-seam die-block for vertical lead cable press

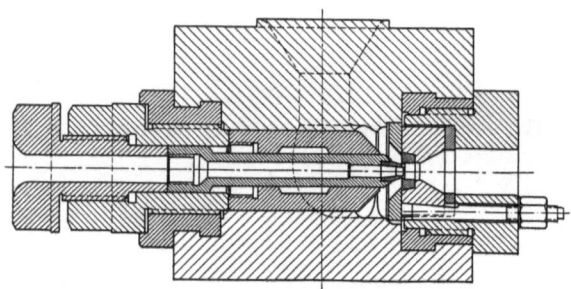

Fig. 20. Split two-seam die-block, as per Fig. 19; however, with tooling for small diameter cables

two lateral oval-shaped lead inlets c to the center bore, which accommodates on the mandrel side the mandrel-holder d with exchangeable point e, the screwed joint f, the one-part threaded bush g and the protecting sleeve h. On the die-side there are arranged the die i with base ring j in front of the former, the die-holder k with four centering screws l and the thrust collar m, as well as the screwed joint n with the one-part threaded bush o. Bushes g and o are to prevent remachining in the block in the event of a thread being demaged. Base ring j serves for the fine regulation of the flow of lead for a given diameter range of the die. The inlet orifice is of funnel-shaped, oval design which is often corrected by lengthy redressing. Coarse regulation of the flow of lead is obtained by a nose p in the center bore, of which part of its bottom circumference is machined off to cater for a reduction of the flow resistance which, on account of the longer distance, is increased. Fig. 21 shows a pipe extruded while adjusting the tools; during this operation the flow of lead towards the upper part of the sheath must be increased by correcting the base ring, so as to eliminate wrinkling on the bottom part of the sheath.

Having established the correct flow of lead, a pipe is first extruded and its wall thickness gauged. Any undue variations may be obviated by radially adjusting the die with the help of the centering screws. The wall thickness of the sheathing is obtained by axially adjusting the mandrel-point and the mandrel-holder respectively, with the help of the rear screwed joint. However, this adjustment is limited and depends on the angle α of the mandrel point. If, for example, the position of its fore-edge in the die is as shown by the dotted line in Fig. 22, the extruded pipe will be too wide, whereas if the fore-edge of the point is too far in front of the die (dash-and-dot line), the sheath will squeeze the cable core too heavily. Experience has proved that the optimum

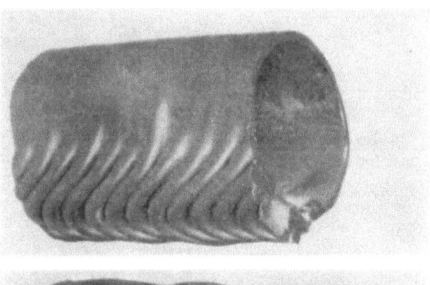
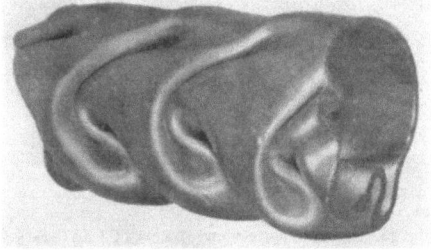
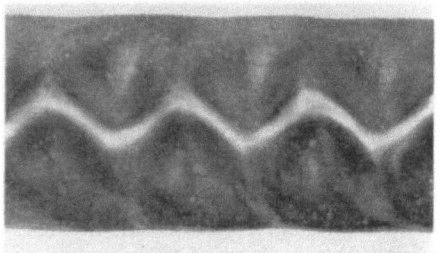

Fig. 21. Sheathing with wrinkling. Trial extrusions with uncorrected base ring prior to inserting cable core

position of the point is in the die plane or slightly in front of it. The flow of lead is schematically shown in Fig. 23.

The concentricity of the wall thickness over the full length of the cable is largely affected by the correct seating of the mandrel-holder; care must be taken that the point cannot alter its position, i. e. deflect elastically, and for this reason it must be supported as close to the die as possible.

In diameter cables most commonly range from 5 to 120 mm. If the production schedule does not require a frequent change of the tooling.

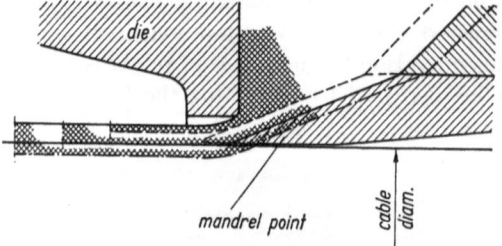

Fig. 22. Positions of mandrel in the die

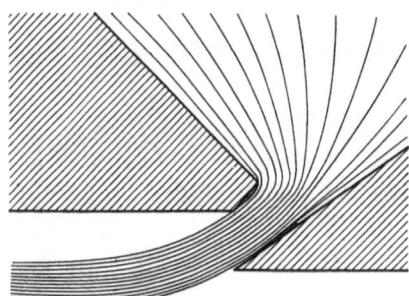

Fig. 23. Diagrammatic view of flow of lead between mandrel and die

it is advisable to use a smaller die-
block for the lower range. It is also
possible, as shown by Fig. 20, to
employ special tooling insets for the
smaller cables which will, however,
entail an increase in the number of
component parts thus contributing
– on account of the clearances required – to a further increase in the
sources of error.

The container, also called recei-
ver, is centered with its inner bore
on the projection of the die-block. It
consists of a heavy jacket into which
a liner is shrunk, the latter being
provided with a slight taper over its

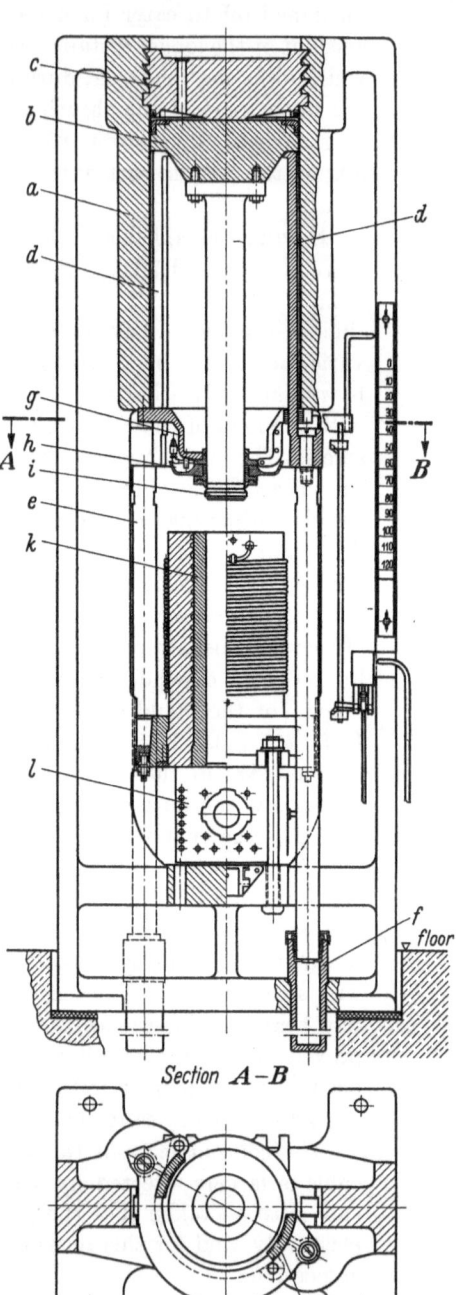

Section A–B

Fig. 24. 3,000-ton vertical lead cable press in frame
construction, with hydraulic cylinder arranged in
the head of the frame, with disk piston, moving
extrusion ram and stationary container (By Krupp
Grusonwerk, Magdeburg)

outside diameter so as to facilitate mounting of this relatively long part. After longer stoppages, the container must be heated in order to obviate thermal stresses and during operation it must be cooled so that the liquid lead solidifies quickly after charging. For this purpose, as illustrated in Fig. 24 and 27, the inside bore of the jacket is provided with a helical

groove, which forms a channel with the wall of the
liner, through which hot
water is circulated for cooling, and steam for heating.
Datas regarding the calculation of containers are
shown on p. 166.

In the course of time
some inconveniences were
experienced which were due
to the movability of die-
block and container. These
appeared at the feeding-
and outlet pipes for steam
and cooling water which
were made up of hoses or
telescopic tubes. Leaks often occurred as a conse-
quence of the permanent
up-and-down-motion.   In
many cases gas hoses are
used for the heating of the
die-block which give rise to
more trouble. With the extension of works facilities
more spacious factory build-
ings were constructed generally and therefore less
objections were raised with

Fig. 25. Shop photo of press shown in Fig. 24

regard to the headroom required by the presses. Thus the down-acting press gradually found more and more application. In this type of press the ram is moving, while die-block and container are stationary so that all of the piping required for heating and cooling of the tools is fixed. Protection of the cable against leaking water was obtained by means of appropriately designed stuffing boxes.

Like up-acting presses, these down-acting presses are in frame- or column construction. Fig. 24, 25 and 26 illustrate the design of a frame-type press. In its head part, as shown in Fig. 24, there is arranged the main cylinder a which is integrally cast with the frame. It is lined with a bronze bush in which the disk piston b moves. The disk piston is sealed by a packing collar. Changing of the latter is carried out after removing the cylinder bottom c which is also sealed by a packing collar and provided with a bayonet union. A disk piston being used, it may be fully accommodated in the cylinder, thus reducing headroom and weight of the press; however, sufficient free space must be provided over the cylinder to allow for the removal of the cylinder bottom. The return motion of the disk piston is performed by means of two segment bars d, engaging on the bottom side of the disk piston and being connected to two return plungers e which are arranged diagonally opposite to them. The return cylinders f are sunk in a pit in the foundation and are rigidly inserted in the base plate of the frame.

Fig. 26. Shop photo of a vertical 3,000-ton lead cable press in frame construction, as per Fig. 24; however, with plunger piston (By Fried. Krupp, Maschinen- u. Stahlbau, Rheinhausen)

The ram is guided in its travel by a cap g, attached to the underside of the cylinder and provided with a gland and an oil lubrication for the ram. Water dripping from the disk piston is collected in a drip pan h and drained off; consequently, it cannot flow down the ram. The ram-head i moves in the container k with a clearance of about 0.75 mm; container and die-block are accurately aligned by means of adjustable wedges on the frame.

In the two-column press, Fig. 27 and 28, and the frame-type press, Fig. 26, a plunger piston has been chosen instead of a disk piston. Although

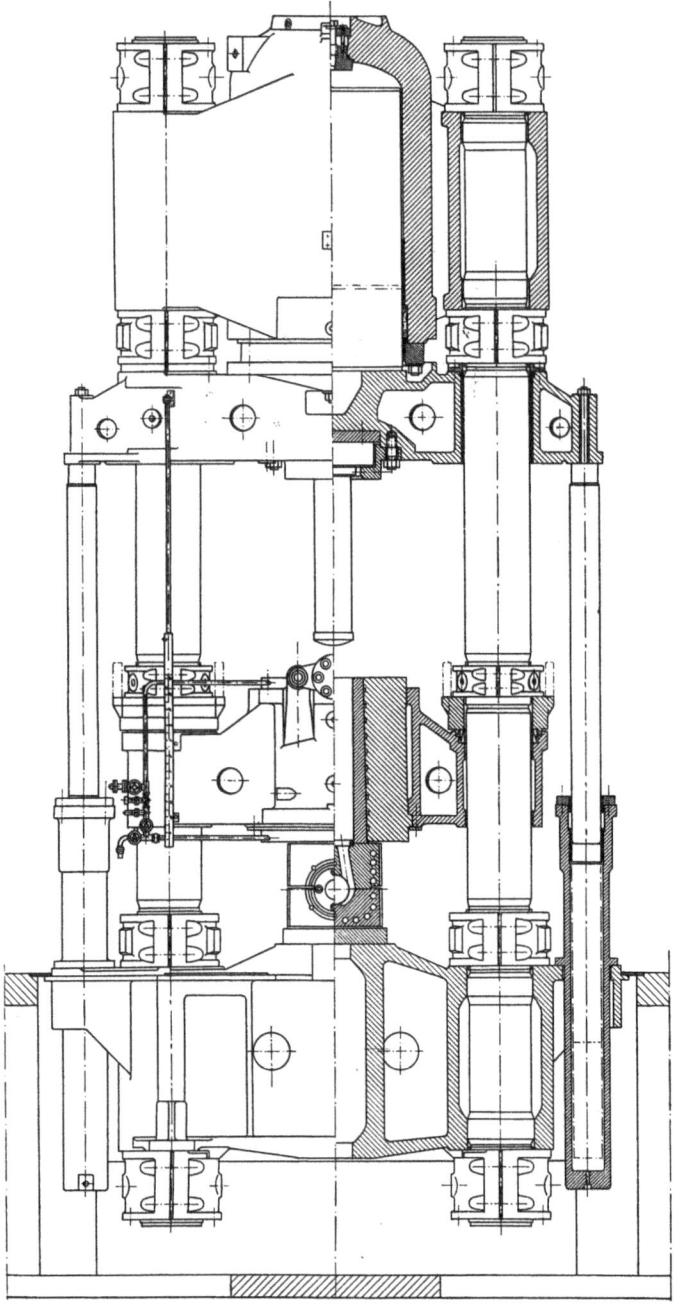

Fig. 27. 2,000-ton vertical lead cable press in two-column design, with hydraulic cylinder arranged
in the top, with plunger piston, moving extrusion ram and stationary container

headroom and weight of the press as compared with the frame-type press just described, are increased, this design has the advantage that changing of packings is simplified considerably. In this case sealing is not effected by a simple packing-collar but by a stuffing box containing a packing which is simply retightened when leakages occur, and a chamber to collect the leak water.

Fig. 28. Shop photo Nordenham of press, as per Fig. 27 (By Hydraulik, Duisburg)

The two-column press, as shown in Fig. 27 and 28, takes the same form of a closed frame as the up-acting press, in that the frame is formed by the cylinder-platen, the bottom platen and the two columns, the ends of which are prestressed by nuts and check nuts on the platens. The plunger crosshead is guided in its travel over the columns and is moved up by two lateral return plungers. The return cylinders are inserted in the bottom platen. The extrusion ram is attached to the plunger crosshead and may be radially adjusted by means of centering screws. The container is set in a crosshead and is forced on to the die-block by nuts on the columns, thus doing away with special tie-rods, as shown in Fig. 24 and 25. Under the column nuts are arranged annular springs serving to hold container and die-block under their prestress during the return motion, too; the down-acting spring pressure is consequently higher than the maximum return power.

Time-tried dimensions of this type of cable sheathing press are compiled in Table 3.

Table 3. Vertical Lead Cable Presses

Pressure capacity, tons	1,400	2,000	2,800
Effective lead charge, kg	300	450	600
Container diameter, mm	180	215	255
Ram pressure, kg/cm²	5,500	5,500	5,500
Ram speed, mm/sec	200	180	170
Pump capacity, H. P.	80	100	140
Max. operating pressure, atm	400	400	400
Pump delivery, l/min	70	90	120
Contents of melting-pot, kg	1,500	2,000	2,000
Average number of cycles per hour	3	2.5	2.2
Max. cable diameter, mm	75	125	125

Development was not limited to the design of the press proper; great importance was attached to improving the tooling – and in particular to the design of the die-block – in order to meet with the increasingly high demands on the cable sheathing. This work called for wide practical experience and a close study of the flow of lead[1]).

For the purpose of investigating flow phenomena the container of a cable press capable of holding a charge of 600 kg, as shown in Fig. 28, was filled with 60 disks each having a thickness of 20 mm and a diameter of 230 mm, alternately of pure lead and lead alloy containing 0.2% antimony, and this charge was then extruded. The hollow space in the die-block had previously been run in with the same alloy. The emerging lead pipe had an outside diameter of 44.5 mm, a wall thickness of 3.3 mm and

---

[1]) GÖLER, V., u. SCHMID: Über die Fließvorgänge beim Pressen von Kabelmänteln. Z. Metallkde. 1939, H. 3.

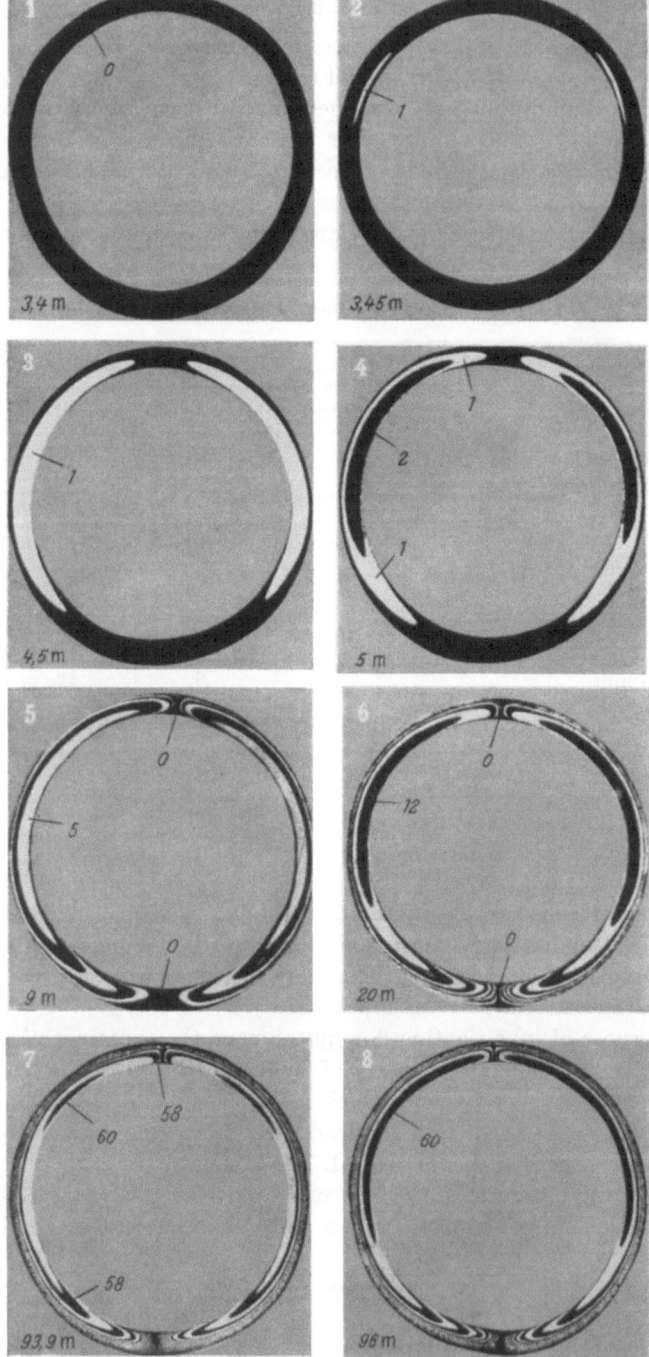

Fig. 29. Sections of lead cable sheath cut at various lengths from the die

a length of 108 m. Extrusion was carried out at a speed rate of 6 m/min and a temperature of about 170 °C.

The lead pipe was then cut at several positions. Fig. 29 shows sections of the pipe at 8 points cut at different distances from the die. Traces of the first disk of pure lead can be seen at *2* at a pipe length of 3.45 m. These have increased to two sickles at *3*. In the following figures more

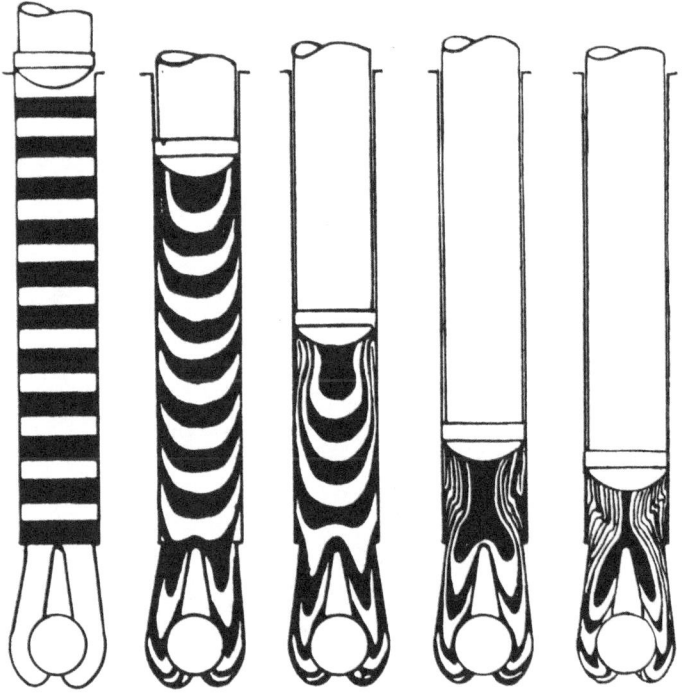

Fig. 30. Reconstruction of flow of lead in container and die-block during experiment, as per Fig. 29

and more disks intrude into the section in the form of sickles and in the last figure they appear in the shape of bark.

The flow lines in the hollow space of the die-block, as illustrated in Fig. 30, were derived from the transverse rings cut from the lead pipe, as it was impossible to remove the solid mass of lead from the die-block without damaging it. However, observations on the flow process in the die-block are available from laboratory experiments carried out with plasticine, the behavior of which is, however, only approximate to that of lead. Fig. 30 shows the extrusion ram in five different positions.

In a normal lead sheath, as illustrated in Fig. 31, the tongues may in most cases be recognized only by slight inclusions of oxides in the two seams. These seams – though not impairing the strength of the sheath – are liable to be the site of inclusions of oxides and impurities in the lead which prevent the intergrowth of the crystals, thus leading easily to the formation of splits or laminations in the sheath, as shown in Fig. 32.

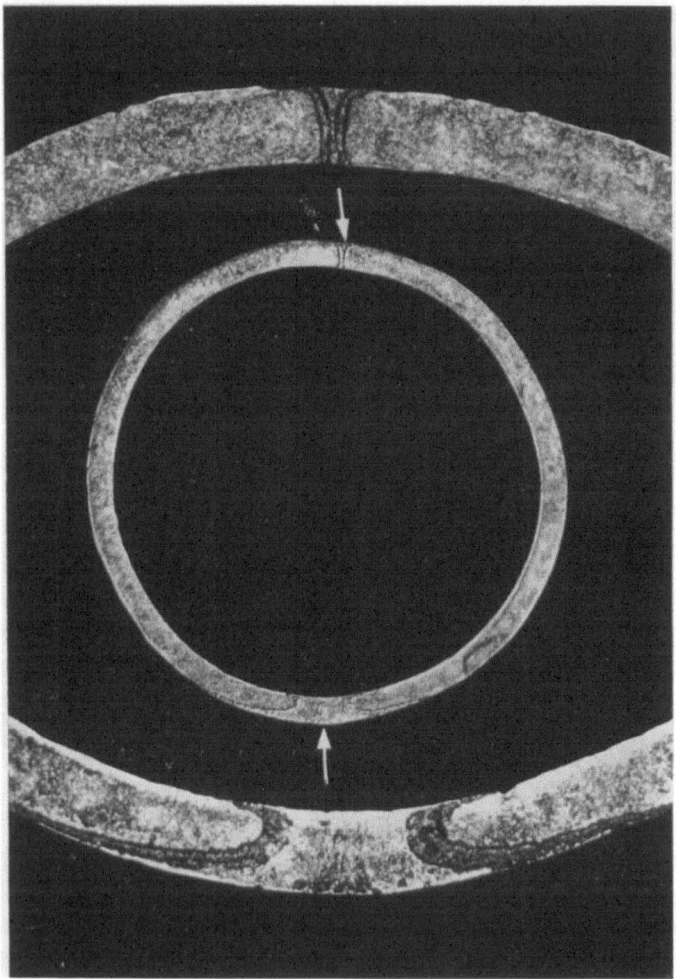

Fig. 31. Seam in lead sheath extruded with tooling, as per Fig. 19 and 20

Efforts concentrated on improving the flow process lead to the development of the single-seam die-block which differs from the older die-blocks by its improved lead inlet. While in the older die-blocks and also in the two-ram presses the lead flows in two separate streams to the chamber in front of the die, it enters this chamber in the single-seam die-blocks in one stream.

Fig. 33 and 34 illustrate a single-seam die-block in which the two inlet bores in the top part are simply connected by a port which prevents

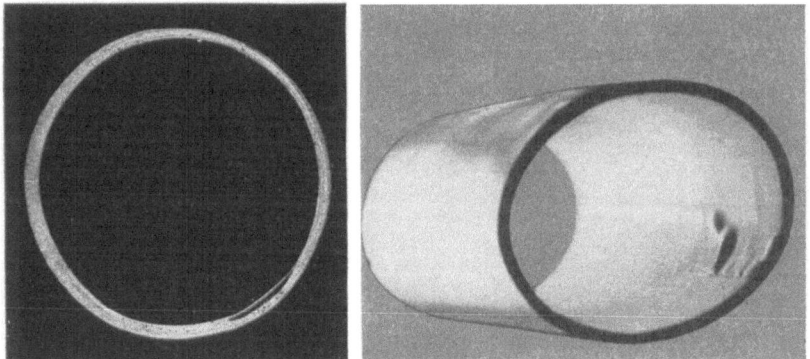

Fig. 32. Defects in lead sheaths

the formation of the top seam. The width of the port is derived from experience.

Fig. 35 shows a section of a sheath having one seam only as it results from properly dimensioned inlet cross-sections.

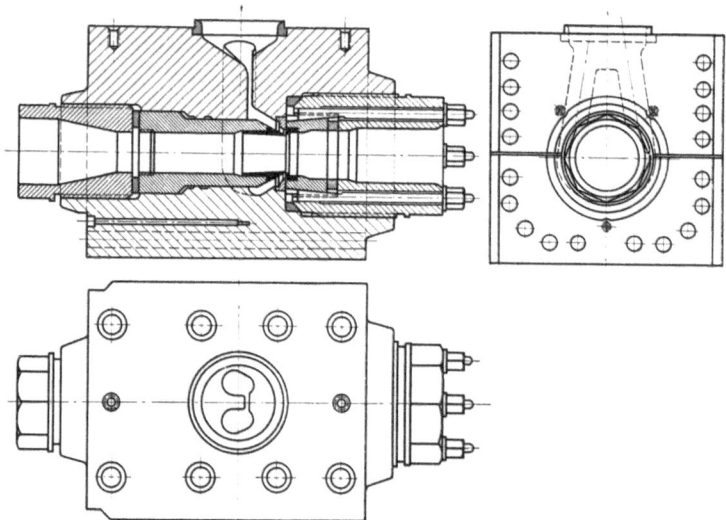

Fig. 33. Split single-seam die-block

Splitting of the die-block, Fig. 34, through its median plane is necessary to enable rectification of the transition points in the lead chan-

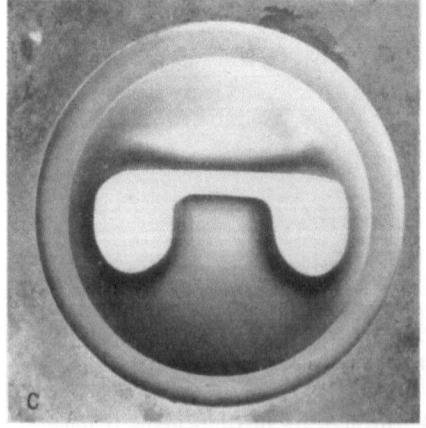

nels. Thus also a long support for the mandrel or the mandrel-holder is obtained which contributes to prevent deflection of the mandrel-point when pressure is applied to or released from the die-block. Variations in the wall thickness of the sheath are thus reduced.

Fig. 36 illustrates a one-part die-block in which the two inlet bores are not connected by a port, but over their full length and take the form of an oblong hole. Regulation of flow is achieved by means of an exchangeable ring which is provided in its top part with a throttling shoulder. This shoulder prevents that too much lead flows into the top half of the sheath. Furthermore the ring has a collar in its inner bore which serves to block the lead and to balance the speed of flow. Installation of this ring, however, necessitates an additional screwed joint in the die-block. A bolt is provided in the bottom of the die-block on which the mandrel is supported. The mandrel screw fitting is in turn inserted in a bayonet union so as to reduce the specific surface pressure on the pitches in the die-block.

Fig. 37 shows the same die-block with tooling inserts for smallest cable diameters.

Defects in cable sheathing due to impurities, inclusions of dross, heavy accumulation of

Fig. 34 a–c. Shop photo of split single-seam die-block (By Hydraulik, Duisburg)

oxides, air- or gas bubbles in the lead charge, may cause great expenses, especially when they are not noticed until the cable has been laid. Greatest care is therefore taken when filling the container and it is for

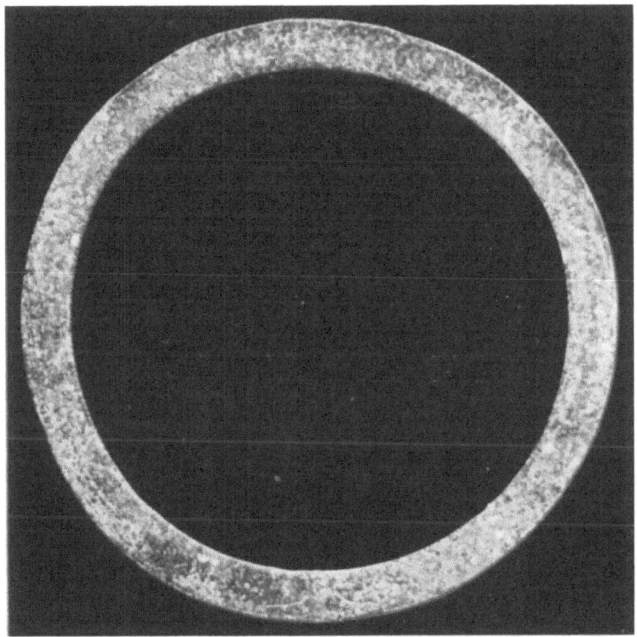

Fig. 35. Seam in lead sheath extruded with tooling, as per Fig. 33

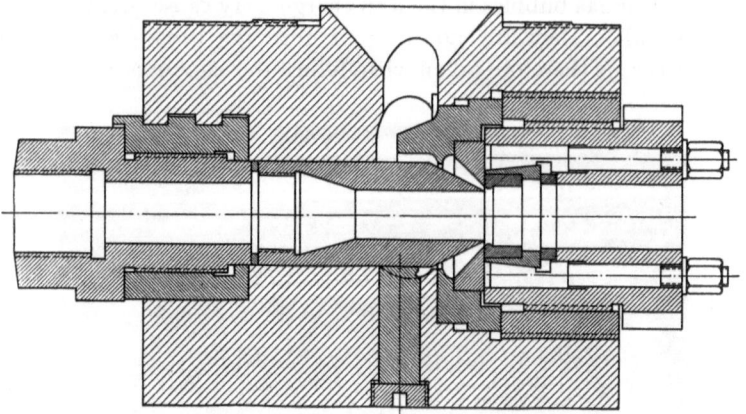

Fig. 36. One-piece single-seam die-block (By Krupp Grusonwerk, Magdeburg)

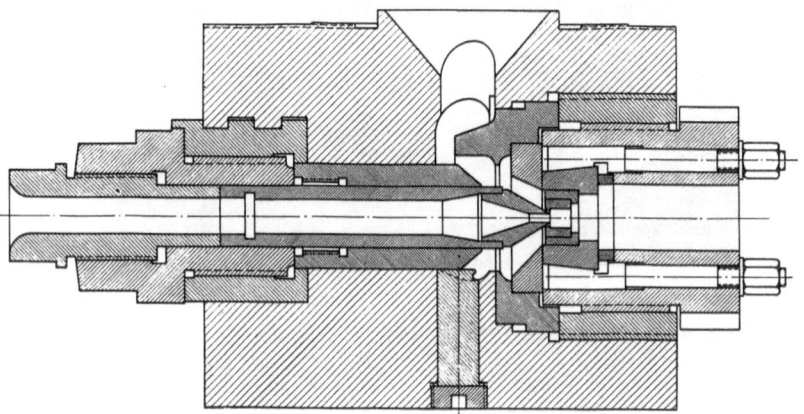

Fig. 37. Die-block, as per Fig. 36; however, with tooling for small cable diameters

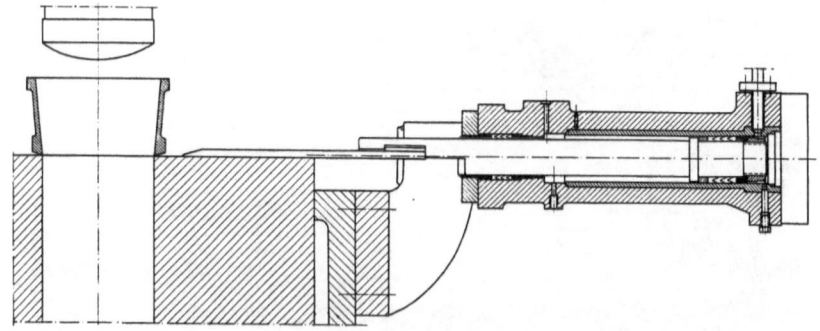

Fig. 38. Shears on container of a vertical lead cable press for cutting of pouring hopper

this reason that pouring of liquid lead is preferred to charging of solid, cylindrical lead billets. Oxidation while filling the container is reduced by

feeding the liquid lead from the melting furnace through a pipe or hose filled with protective gas, and pouring being carried out from the bottom. Prior to pouring a high funnel, in which dross and foreign particles may collect, is placed on the container. To prevent a contraction pipe forming in the lead as it solidifies, the extrusion ram is brought under slight pressure, i. e. under the load exerted on it by the weight of the moving parts, into the filling funnel so that the lead solidifies under pressure. A period of 10 to 15 minutes is allowed for freezing and the extrusion ram is then moved down slightly thus causing the tapered funnel to rise and to release a slit for shearing. The lead residue is ultimately knocked out of the funnel and remelted.

The operating cylinder of the shear which is to separate the funnel level with the container top, has a double-acting piston and is mounted in a bracket attached to the container-

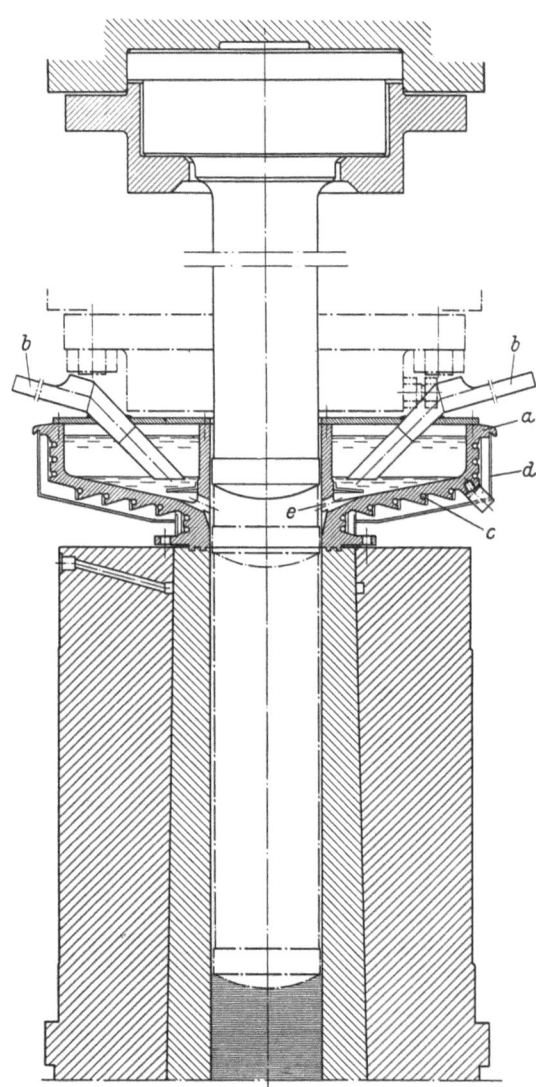

Fig. 39. Equipment for pouring molten lead into container in vacuo (Glover & Co., Manchester)

holder, as may be seen in Fig. 38. The wide shearing blade of about 10 mm thickness is screwed to the piston rod which is forged down in the form of a spade, and is guided in its travel on the container in lateral gibs.

Oxidation chiefly occurs after completion of the extrusion and after
the ram has been retracted and is caused by the air entering the container.
The oxides mix with the lead crystals and do not cause any harm as long
as they are only finely dispersed in the structure. Oxidation and occlusion

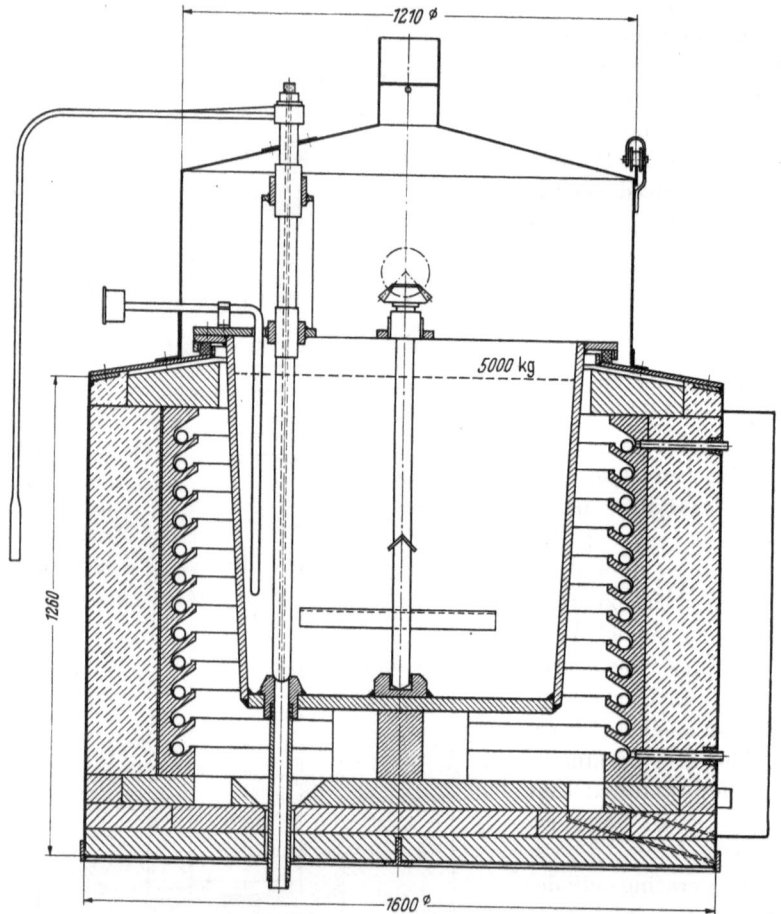

Fig. 40. Electrically heated lead melting furnace of 5,000 kg contents

of harmful air-bubbles can be greatly eliminated by pouring the lead in
an evacuated container. An example of this is illustrated in Fig. 39, in
which a vessel a, to which the lead is supplied via pipes b, is placed on the
container. The vessel is heated by electric heating coils c on its outside.
An insulating casing d is provided to prevent heat radiation. In the bottom of the vessel there is a number of orifices arranged at the cylindrical

ram guide, through which the lead passes to the ram during the extrusion, thus sealing it against the atmospheric air. When the ram is retracted from the container upon completion of the extrusion cycle until it comes clear of the orifices, the liquid lead flows into a vacuous space and the layer of lead on the ram head maintains the vacuum seal.

Experiments have been made with various other suggestions to evacuate the container, none of them having, however, given satisfactory results in practice. Vacuum casting has in general not found a wide application and preference is given to the simple charging method using a funnel.

The old kettle in use for the melting of the lead was exclusively of the coke-, gas- or oil-fired type, while nearly all of the newer melting units are electrically heated. An example of the modern type is illustrated in Fig. 40. The melting kettle consists of a large pot made from steel plate, placed in a heating chamber and covered with a sheet-metal hood. The heating chamber is built from refractory material and enclosed, on its outside, by a sheet-metal casing. The heating wires, strips or cartridges are arranged on its cylindrical inner wall. The pot is provided in its center with a bevel-gear driven stirrer and on its side with a spindle by which the lead outlet in the bottom is opened or closed. An immersion tube containing a thermocouple to control the temperature of the lead, is arranged close to the spindle. The stirrer serves to mix lead and alloying metals. The pig lead is charged through a lateral feeder door in the hood. The waste gases are exhausted through the upper duct. Rectangular vats are frequently used instead of the round pots.

Tongues and seams in the cable sheath which – although they do not affect its strength – are more liable than any other place in the cross-section of the sheath to be the site of oxidations, are caused by the fact that the ram is disposed perpendicular to the axis of the cable. When the two axes coincide, the seams will disappear as the flow will now be uniform in the cross-section of the sheath.

A horizontal press of this type is shown in Fig. 41. The lead is charged in a container disposed concentrically to the mandrel and the mandrel-holder respectively, and after solidification it is forced through the die in a manner similar to that used in a lead pipe press, see p. 4.

The press has a cylinder platen a being positively connected by four columns b to the counterplaten, on which is arranged the container d with tooling. The base ring e is rigidly mounted in the bottom of the container. Die-holder f and die g are located behind the base ring. The die may be adjusted by means of the screw-fitting h in the mandrel axis thus rendering possible a variation of the outside diameter of the cable. Radial positioning of the die-holder f is achieved by wedge bolts. Alternate tightening and releasing of the wedge bolts causes the die to be adjusted radially so that it may be centered to the axis of the mandrel. Thus a uniform wall thickness in the cross-section of the cable is achieved. Compressive stresses on the die-holder and the die are taken up by the screw-fitting k.

The tip of mandrel l is seated in a mandrel-holder m of which one end is provided with a threaded flange n and screwed into the container

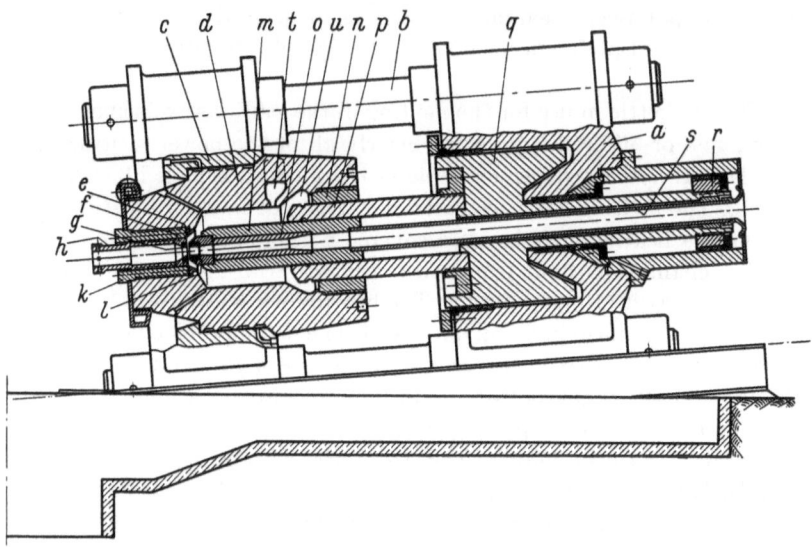

Fig. 41. 2,650-ton horizontal lead cable press with equiaxial arrangement of ram and die
(By Henleys Telegraph Works, London)

orifice. The tip is additionally supported in the mandrel-holder by a tube o so as to keep to a minimum the deflection of the tip and the variation in the wall thickness respectively, when releasing and retracting the ram p. The ram is provided with four elongated slots to allow for the passage of the stems which connect the mandrel-holder m to its flange n. The ram pressure is produced by the main plunger q, the return motion of which being effected by a piston r on the plunger extension. The drawback cylinder is bolted to the base of the cylinder-platen. The cable enters through a protection tube s which extends through the bore of the main plunger and is fixed in the mandrel-holder.

The container is charged with molten lead through the orifice t when the ram head is slightly advanced so as to close tightly the container bore. The press is set on its foundation so that its front end is inclined down so as to permit the complete escape of air from the container while it is being filled; the inclination is $1:12$. A chamber u which is easily accessible from the side of the press, is arranged in the container flange in front of the filling hole so as to permit the cleaning of the ram and the removal of shell residues.

When changig the mandrel, the lead slug behind the die must be removed. This is done by turning the screw-fitting h completely out of the thread and dismounting the die towards the rear. As may be seen in Fig. 42, the lead slug is milled out by a cutter-head v set on a tube w and advanced and rotated by a motor, the worm drive x and the threaded ring y.

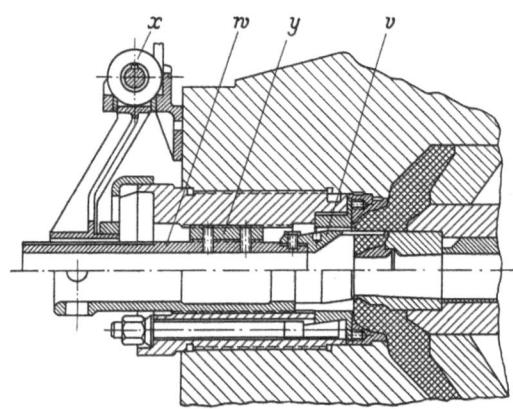

The press capacity is 2,650 tons. It takes a lead charge of 450 kg at a container diameter of 400 mm; its mandrel-holder diameter is 240 mm and its effective stroke 485 mm. The extrusion ram pressure is approx. 3,250 kg/cm².

Fig. 42. Tool for removal of lead slug in front of the die in the press, as per Fig. 41

The outstanding features of this new design could not make up for the disadvantages encountered in charging the press and the long distance to be covered by the entering cable. It is for this reason that the press has not found any application worth mentioning.

With vertical single-ram presses calculation of the press tonnage is based upon a ram pressure of about 5,500 kg/cm² which takes account of lead alloys containing up to 1% antimony can be extruded, too. For the extrusion of pure lead a ram pressure of about 3,800 kg/cm² would be sufficient. However, this lower pressure is employed in relatively few cases only, as lead alloys have to be extruded on the presses most frequently and changing over to a larger container and ram occupies too much time. The charge of the container is given by the ratio $l : d \approx 1 : 5$, in which l is the filling height and d the container diameter.

During extrusion the temperature of the tools and the die-block is to be brought to about 180 to 200 °C. Thermocouples arranged at suitable positions in the top- and bottom part of the die-block, provide the necessary indication for control. It is advantageous to drill longitudinal holes in the die-block to accomodate the electric cartridge heaters and to divide the area to be heated into various zones which are individually controlled. This is important and necessary to ensure an irreproachable flow of the lead.

It is only when starting the operation of the press that the container is heated up, this being done as a rule by steam, to a temperature of about 100 °C. The molten lead must not be poured into a cold container as this would cause cracks in the steel. Prior to and during the solidification period the container is cooled with steam. If the heat elemination is not sufficient, hot water is passed through the steam ducts; the temperature of the water must not be below 80 °C to avoid thermal stresses.

If the press is driven by pressure water, it is general practice to install two horizontal or vertical three-piston pressure pumps (see p. 204) so that in the event of repairs the press may be operated with one pump only. The delivery volume of the pumps depends on the cable speed, which in turn is dependent on the cable diameter and the alloying metals in the lead. The speed range is about $v = 5$ to 60 m/min according to the maximum and minimum cable diameters. When the lead, however, contains antimony or other alloying metals, this speed is essentially reduced. Table 3 shows pump capacities having given good results in operation.

Having calculated the maximum ram speed v_1 from the cable velocity v, the delivery volume of the pumps is given by $Q = f \times v_1$ in which f is the piston area of the main plunger. The return speed v_R of the ram may be assumed $v_R = 10\, v_1$, because it is general practice to rate the return power of the press at 10% of the press tonnage – an empirical value chiefly depending upon the friction of the ram in the container.

The ram speed must be regulable to suit the different cable speeds. This is done by regulating the delivery volume of the pumps by means of a throttle valve. It is more advantageous and economic, however, to have one of the pumps driven by a variable-speed motor.

The production capacity of a cable press depends largely upon the ram speeds and therefore varies in a wide range. Allowing 10 to 12 minutes for the solidification of the liquid lead, 3 to 4 extrusion cycles are performed per hour at an average.

Oil-hydraulic pressure pumps are frequently employed for the operation of cable presses instead of pressure water pumps, see p. 203. Their advantage being that they require less floor space and that they may even be mounted on the press. Claims in favor of pressure water pumps are the higher operating pressures, resulting in a simpler design of the press, and, last not least, the elimination of the danger of fire.

Fig. 43 illustrates the hydraulic circuit diagram of a water-hydraulic cable press. The valve housing accomodates the inlet valves *1* and *3* for the drawback- and main cylinders with corresponding outlet valves *2* and *4*, a throttle valve, a control slide valve and the check valve *5*. Valve settings in the control positions *extrusion, stop* and *return* are shown in the valve lift diagram. The throttle valve serves to control the pressing speed and to relieve the valves when changing over from one control position to another. With the throttle valve being open, the pumps are idling, in which case the check valve 5 prevents the ram from moving down. The control slide valve actuates the shear and sets it to *shearing* or *return*.

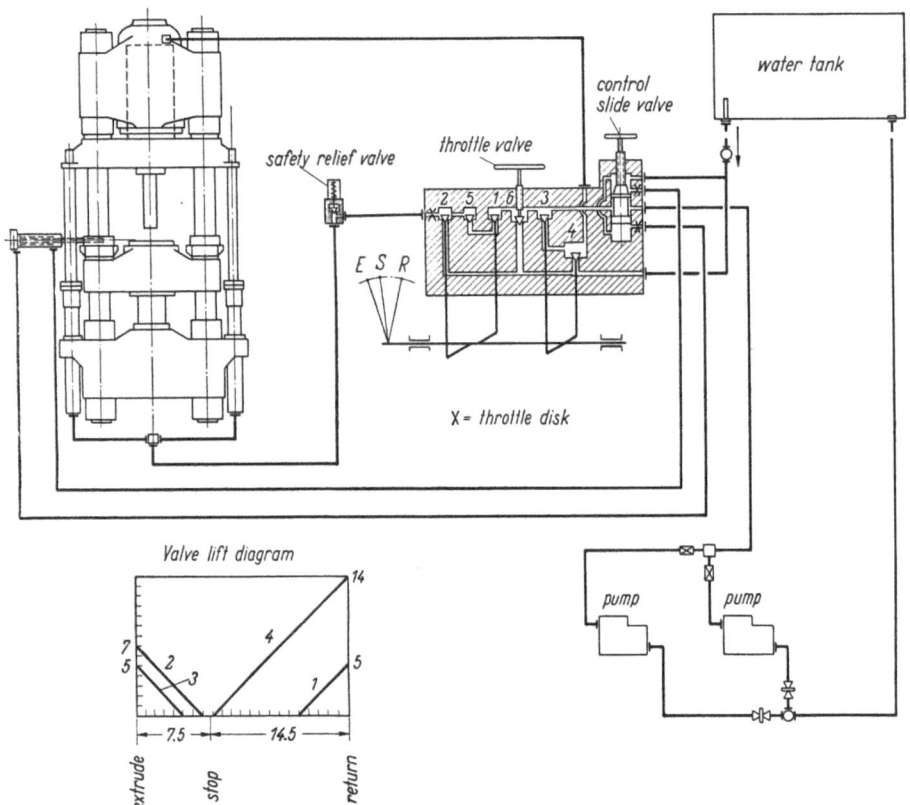

Fig. 43. Hydraulic circuit diagram of a vertical lead cable press with direct pump drive, as per Fig. 27

b) Lead- and Aluminum Cable Presses for Billet Extrusion

In the manufacture of rubber-insulated cables or rubber hoses the lead sheath is applied for the vulcanizing process only. The sheathed cable or hose is coiled on drums which are placed in a steam-heated tank which is closed tightly with a lid. During vulcanization the hoses must be forced from their inside against the lead sheath. To achieve this, they are filled with hot water which at the vulcanizing temperature of 140 to 150 °C attains a pressure of 4 to 5 atm. Upon completion of the vulcanization, the cable or hose is wound off the drum and passed through a stripping- and chopping machine, in which the lead sheath is ripped open, stripped off and chopped in one operation. The chopped lead is then returned to the lead melting kettle.

The vulcanized hose retains the curvature of the drum after having been wound off and that is why this method is rarely employed in Germany at the present time. The method preferred now is that in which

Fig. 44. Shop photo of 2,000-ton vertical lead sheathing press as used for the vulcanization of cables. Press in four-column design with underlying hydraulic cylinder, moving container, traversable extrusion ram and swinging device for charging of billets (By Breda Fucine, Milan)

straight lengths of rubber hose of about 30 m are lead sheathed and then placed in vulcanizing tanks of suitable length.

Requirements to be met by this kind of lead sheathing are very low as compared with lead sheathing as used for cables. In order to eliminate defects, the sheathing is furthermore extruded with a greater wall thickness as is usual for cable sheathing, as the lead is remelted. Should there be any defects, though, these may easily be remedied by soldering.

On account of what has been said in the foregoing, these presses destined for the production of rubber-insulated cable and rubber hoses, may be charged with solid cylindrical lead billets instead of molten lead. Thus the solidification periods are eliminated and a substantial increase in production is achieved.

Fig. 44 illustrates a billet-fed lead cable press. The up-stroking design, as shown in Fig. 18, is the most suitable type, because the ram must be traversed to permit of charging the billet, for which purpose a small hydraulic cylinder with plunger and guide-frame is fixed on the face of the upper platen. With a down-stroking press, this drive unit would have to be mounted on the movable crosshead which would call for connection by hoses or telescopic tubes.

The press of about 2,000 tons capacity is capable of extruding billets of 250 mm diameter and 1,100 mm length. The sequence of the charging operations may be automated by very simple means. A belt conveyor feeds the lead pigs into the melting kettle. The molten lead flows into two cylindrical molds from which the billets are alternately ejected through the top by a hydraulic ram. The billet is then placed in the container by a gripper which pivots with its arm round a column mounted at the side of the press. All of these operations are performed electrically and initiated from a control desk.

Aluminum cable sheathing presses are also charged with solid billets, because pure aluminum shows excellent welding qualities under pressure and the beginning of a new charge can therefore not be traced in the sheathing. Another reason for the use of precast billets is the high melting temperature of the aluminum; the container steels would not be able to withstand the high thermal stresses if molten aluminum were used.

It has only been during the past ten years, that perfect results have been obtained with aluminum cable sheathing presses. Fig. 45 shows a press of the well-known down-stroking type. The container is charged with billets of 200 mm diameter and about 1,000 mm length. They are continuously cast from aluminum of 99.5 to 99.8% purity and heated from room temperature to about 450 °C in an induction furnace located close to the press. After a heating period of about 10 to 15 minutes a hydraulic ram pushes the billet vertically through the top of the furnace, where it is gripped by a tongs and placed into the container. The ram stroke is dimensioned to provide sufficient space between ram and container to allow for the charging of the billet. The press is capable of sheathing cables ranging in diameter from 6 to 120 mm. The wall thickness varies from 0.3 to 3 mm. The maximum pressure capacity is 3,800 tons, which, however, is only required at the beginning of the extrusion cycle and decreases to about $2/3$ of this value towards the end of the stroke. A relatively high drawback power must be applied to withdraw the ram from the container. It amounts to about 15 to 20% of the press tonnage. Container and die-block are heated and holes are drilled in the longitudinal axis of the container jacket to accommodate resistance heating coils or copper rods for induction heating, see p. 178. The container is initially heated to about 400 °C. Heating is carried out

over a period of a few days in order to avoid major temperature differences which might cause undue thermal stresses in the container. It is

Fig. 45. 3,800-ton vertical aluminum and lead cable press in two-column design, with upper, moving
extrusion ram and fixed container. Courtesy of Siemens-Schuckertwerke, Berlin-Gartenfeld
(By Hydraulik, Duisburg)

for this reason that switching arrangements are often made which automatically cut the heating current as soon as a temperature difference
of about 50 to 100 °C is exceeded in the container.

The tooling – i. e. extrusion ram, container and die-block with screw fittings, mandrels and dies – is made from high temperature, annealed steels having a tensile strength of 120 to 150 kg/mm².

Cooling of the cable requires very close attention. It must prevent any charring of the cable core when the press is stopped for recharging. The cooling pipes on either side of the block therefore extend as close as possible to the mandrel-tip and the die. The cooling pipe on the mandrel side protects the incoming cable core against radiant heat. In most cases it consists of two telescoped tubes and the cooling water is circulated through the concentric annulus formed between them. On the die side a number of small holes is drilled into the inner tube through which the cooling water is sprayed on the sheathed cable. The two tubes are fitted in the respective screw fitting with a little clearance to keep to a minimum the dissipation of heat. The face of the cooling pipe is sealed in the die holder; the cooling water thus flows back from the sprays through the inner tube and not over the hot screw fitting.

In order to minimize the stopping periods, all of the motions required for charging the fresh billet must be performed as rapidly as possible. Time consuming motions are in particular the withdrawal and the lowering of the extrusion ram. It is therefore advantageous to have the return stroke of an aluminum cable sheathing press carried out by pressure water taken from an accumulator. Fig. 46 illustrates a diagram of this method of operation. Control positions and valve motions may be noted from the valve lift diagram. In the stop- and/or charging position all of the valves are closed. When the control lever is shifted to pressure after the billet has been charged, valves 2 and 3 are opened. The ram travels down, the low-pressure water contained in the air-vessel, forces the filling valve open and fills the main cylinder. The idle motion having been completed, the pumps deliver into the cylinder; during this operation the pressure rises to 400 atmospheres. Upon completion of the extrusion cycle, the control lever is shifted to "return". All of the previously opened valves are now closed one by one. The pressure water from the accumulator now passes through the opened valve 1 into the drawback cylinders, to which is also connected the driving cylinder of the filling valve. The latter is forced open when the main cylinder has been released, by the driving piston in that valve 1a is pushed open. The ram travels up and forces the water out of the cylinder into the air-vessel (see p. 216).

Aluminum cable sheathing presses are both of the single-ram and the twin-ram type. The twin-ram press offers the advantage that the resistance to deformation in the die-block is lower, but is not practical for presses in which both aluminum and lead sheathing is extruded. Twin-ram presses are made both in vertical and horizontal design. Choice of type depends on the space available at the site of erection. If space is limited, preference is given to the vertical design, in which case, however, a high headroom is required. When choosing the horizontal type, more floor space will of course be required.

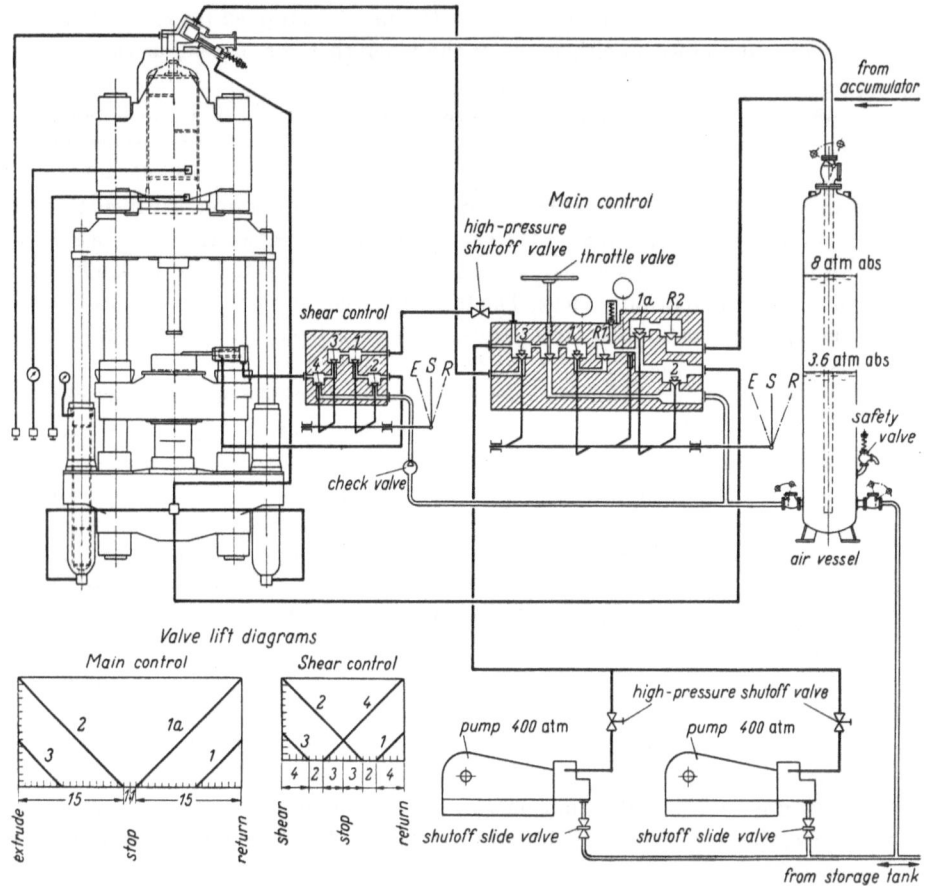

Fig. 46. Hydraulic circuit diagram of cable press, as per Fig. 45

Fig. 47 and 48 illustrate the largest twin-ram press built up to the present time. It is designed for a maximum pressure capacity of 4,500 tons and is capable of simultaneously extruding two billets of 245 mm diameter and 600 mm length each. The total weight of the two billets amounts to 160 kg. This relatively high billet weight permits of sheathing cables of largest diameters. High billet weights increase the capacity of the press and reduce the number of the undesired stop marks.

The center crosshead with the die-block and the two containers is stationary and accomodates the two drawback cylinders a. When the

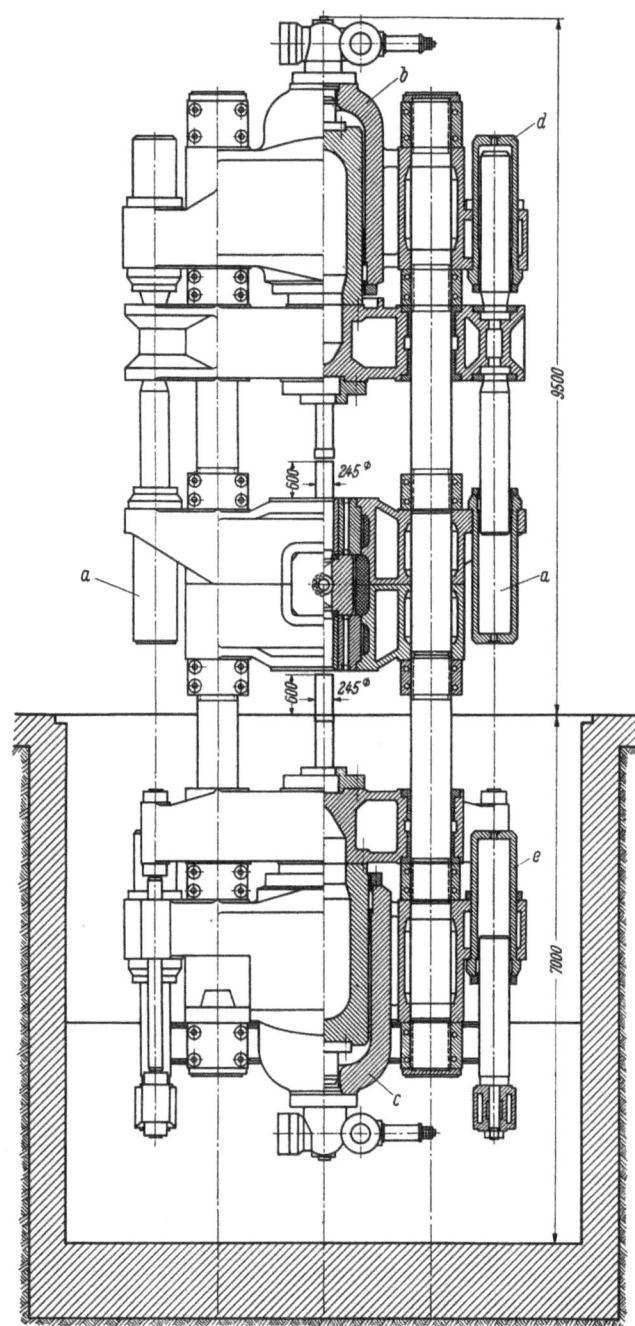

Fig. 47. 4,500-ton
vertical twin-ram
aluminum cable
press in two-col-
umn design with
hydraulic coupling
to ensure positive
ram motion

water is discharged from these cylinders, the pressure of the filling water in the main cylinders b and c causes the top ram to move down and the bottom ram to move up. The billets having been upset, the pumps are switched to deliver into the main cylinders and the connection be-

Fig. 48. Vertical twin-ram aluminum cable press, as per Fig. 47, extruding cables with corrugated sheathing. Courtesy of Felten & Guilleaume Carlswerk AG, Köln-Mülheim (By Hydraulik, Duisburg)

tween the two coupling cylinders d and e is established, while at the same time the drawback cylinders receive a constant counterpressure from the accumulator. Now the two rams are forced to move against each other – this being done under approximately equal pressure – and steadily force the aluminum from both sides towards the die. During this operation the pressure in the main cylinders may rise to 400 atmospheres. For the return stroke this coupling is disengaged. Cylinders a and e are fed with pressure water. Cylinders d and main cylinders b and c are connected to the return piping; the rams move back into their home positions.

Aluminum sheathed cables of more than 60 mm diameter are very stiff and do not lend easily to coiling on a drum. Such difficulties are eliminated by corrugating the sheath directly upon emerging from the die-block[1]). In this case it is also possible to cool the sheath internally

Fig. 49. Shop photo of horizontal 1,600-ton twin-ram aluminum cable press in frame construction
(By Schloemann, Düsseldorf)

by air during the stopping period which permits of insulating the cable core with highly heat-sensitive material.

Fig. 49 illustrates a 1,600-tons horizontal twin-ram press. The two billets have a diameter of 175 mm and a length of 440 mm each, their total weight being about 55 kg. The outside diameter of the largest cable amounts to 60 mm. The billets are heated in the same way as in the presses described in the foregoing, that is in an induction heater, and automatically charged into the containers. In order to evacuate the air from the container, the ends of the billets which are touched by the rams, are cooled by spraying them with water, thus causing a temperature gradient in the billets and upsetting to commence at the hotter end. Con-

[1]) HILGENDORFF, H. J., u. W. THELE: Starkstromkabel mit aufgepreßtem Aluminiummantel. F& G-Mitteilungen 1957. – HAHNE, K. H.: Neue Wege in der Herstellung und Verwendung von Alu-Kabeln. „Aluminium" 28. Jg. (1952) H. 7/8.

tractions in the cable sheath at the stop marks are avoided by hydraulically adjusting the die screw fitting when the die-block is released. Thus the annulus between die and mandrel-tip is slightly increased, due to which the wall thickness increases.

c) Continuous Lead- and Aluminum Cable Sheathing Presses Charged with Molten Metal or Solid Billets

The idea of developing a continuously operating cable sheathing press, by means of which it is possible to eliminate solidification and waiting periods and to avoid stop marks – the so-called "bamboo-rings" –

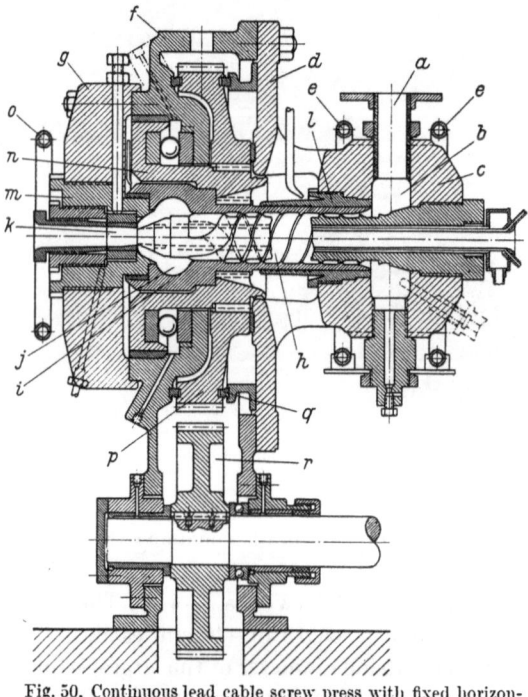

has long been attractive and a study of the relevant patent literature reveals many suggestions how to tackle this problem. Most of these have, however, been unsuccessful or proved to be uneconomic because the expenditure involved was too high.

First success was achieved when a rotating screw was employed to carry molten lead continuously into the die-block, this method being derived from the many screw presses in use for the production of rubber-insulated cable. This development commenced about 30 years ago and eventually lead to the construction of con-

Fig. 50. Continuous lead cable screw press with fixed horizontal screw (By Henley Extrusion Machine, London)

tinuous lead screw presses. Although these machines are beyond the scope of this book, a brief description will be given to round off the picture.

Fig. 50 shows a diagrammatic section of one of the first screw presses. The liquid lead is admitted through the lead inlet *a* to the chamber *b* of the housing *c*, which is provided with a flange *d* and heated by gas burners *e*. The flange is fixed to the gear-box *f*, the opposite side of which is closed by a cover g. The screw is rigidly fixed in the housing c and carries at its end the mandrel-tip. The cable protecting tube in the screw axis is cooled from the outside by air. The screw housing i has a chamber j from where the lead is admitted to the die k. It rotates in the water-

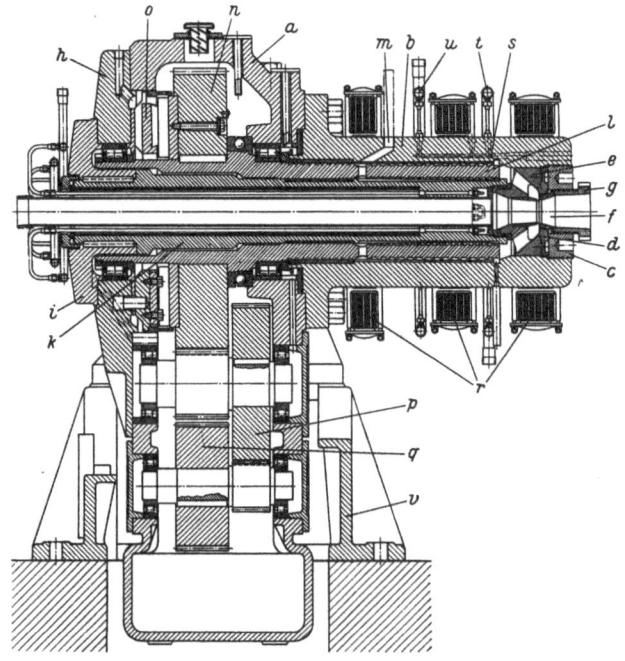

Fig. 51. Continuous lead cable screw press with driven horizontal external and internal screw
(By Pirelli General Cable Works, Southampton)

cooled sleeve l and over the ridge of the nut m which accomodates the screw-fittings for the die and the radially adjustable die-holder. The screw housing rests with sleeve n on a ball bearing and is driven by a spur gear p which prevents leakage of lubricating oil by sealing rings q. The driving spur gear r works on gear p through two lateral intermediate gears. The cover g is heated by the gas-burner o when the temperature drops.

An improved design of this screw press is illustrated in Fig. 51. To the right-hand side of the gear-box a is attached the screw housing b which accomodates in its bottom the adjusting nut c, the die-holder d and the web-shaped tip-holder e. The die f is adjusted and held by the screw-fitting g. The left-hand side of the gear-box is closed by a cover h which carries the mandrel-holder rod k set in a flange i. The rotating and hollow screw casing l is furnished towards its end with external as well as internal screw threads to force forward the lead admitted through m. Labyrinth channels prevent the leakage of liquid lead towards the rear. The screw casing runs in roller bearings. The axial thrust originating from the rotating screw casing is transmitted through the driving spur wheel n to a Mitchell-bearing o. On the opposite side, the spur wheel, which is engaged with two reducing gears p and q, rests on a ball bearing. A double-walled cable protecting tube lies in the mandrel-holder rod. The tube is fitted with an external clearance and is cooled from the inside so as to keep the rod hot and the cable core which passes through it, cold. The screw casing is provided with three induction heating coils. Furthermore it has cooling bores s with manifolds t and u through which the coolant – steam or water – is circulated. The gear-box is mounted on a base frame v. The machine is built in three sizes as per the dimensions tabulated in Table 4.

Table 4. Continuous Lead Cable Screw Presses

Max. Cable O. D. mm	Min Cable I. D. mm	Motor Power H. P.	Lead Displacement kg/min	Max. Heating Wattage kW	Melting Pot Contents kg
26	3	60	11.4	70	2,000
76	9	100	27.3	110	3,500
127	40	150	32.7	135	5,000

Another excellent design of a screw press is shown in Fig. 52. This type makes use of the die-block known from hydraulic cable presses, in which the flow of lead is diverted 90°, while in the previously described screw presses extrusion direction and cable passage lie in the same axis. It will be recalled that the advantage of this kind of die-block is that the mandrel may be adjusted axially and the position of the die may be changed radially, thus permitting of correcting thickness and centering of the sheath during operation. This machine has furthermore a practical, short cable inlet and requires little headroom.

The die-block consists of a cubical block which is fixed on a cylindrical screw housing enclosed by an insulating casing. The solid and slightly tapered screw forces the lead through channels machined in the mandrel-holder, towards the die which is set in a radially adjustable die-ring and held by a screw fitting. Both screw housing and die-block are provided with a resistance heating system and they may also be cooled with steam or hot water in that these coolants are circulated through channels drilled into these parts. The resistor elements are coiled round the screw housing and installed in longitudinal holes drilled into the die-block, whereas the cooling channels are disposed transversely. Thermocouples provide a very accurate means of temperature control.

The driving unit of the screw is installed underground and consists of a gear with a tapered-roller thrust bearing for the screw and two reduction spur gears as well as a variable-speed driving motor. Gear and motor are connected to each other through an overload coupling. The motor

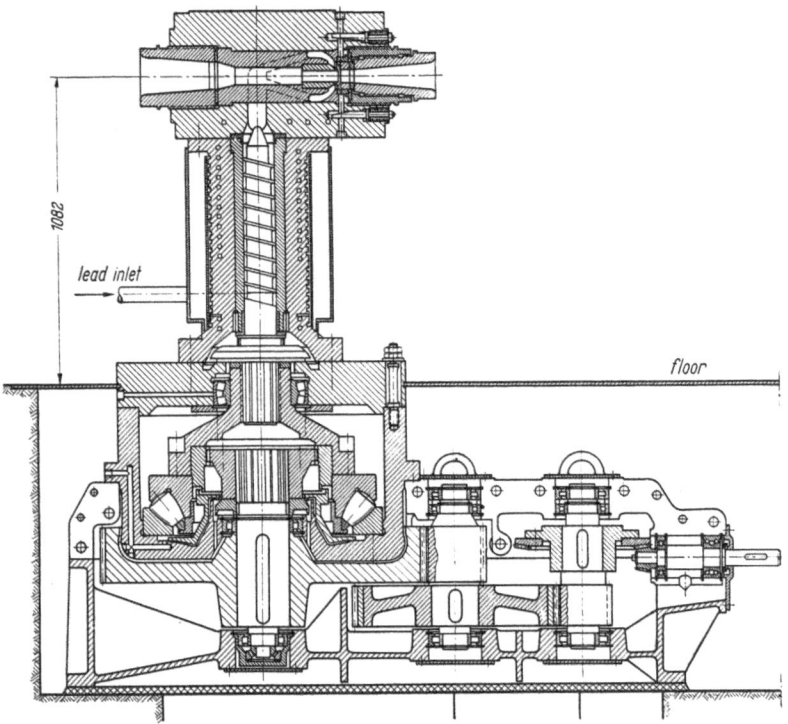

Fig. 52. Continuous lead cable screw press, Hansen System, with driven vertical screw; type Robertson, Brooklyn (By Hydraulik, Duisburg)

power is about 50 kW at $n = 1,450$ r.p.m. The screw rotates at 16 r.p.m. and displaces about 20 kg lead per minute for cables up to about 80 mm dia.

Notwithstanding the many advantages of screw presses, they still have some important imperfections and that is why hydraulic lead cable sheathing presses continue to find a wide application.

A great drawback is that a screw press cannot be suddenly stopped without causing impairing effects, when, for instance, a faulty spot in the insulation of the incoming cable core has to be remedied. In such case temperature troubles causing damage to the sheathing, would occur immediately. Very slowly only may a screw press be stopped or accelerated to full capacity and in doing so, the temperature must be controlled very carefully and this press is therefore suitable for the manufacture of very long cables only without having to change the die or the mandrel. The machine is kept in continuous motion while a filled coiling drum is replaced; the pipe extruded during this time will be remelted. Consequently, the hydraulic press is more suitable when smaller orders are concerned and when the tooling has to be changed frequently.

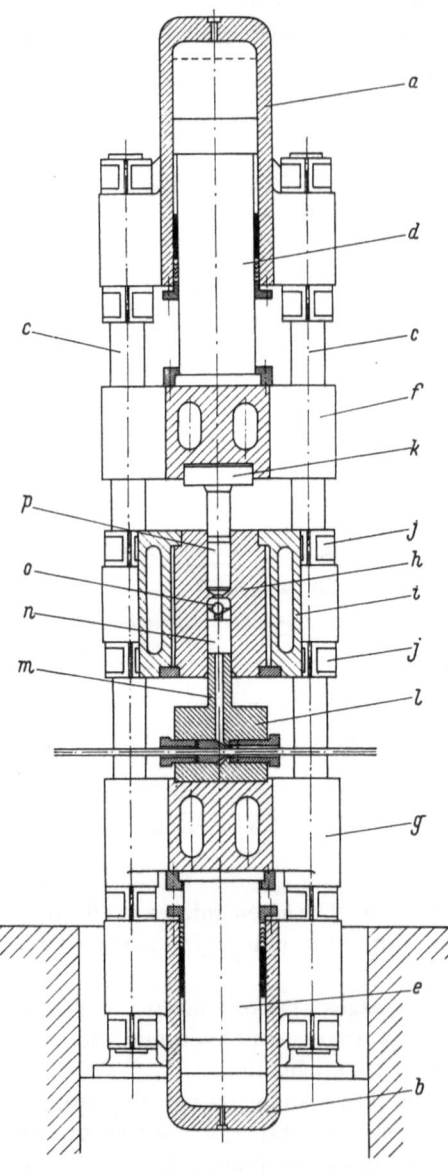

Fig. 53. Continuous twin-ram aluminum cable sheathing press, Hydraulik System

Another drawback is met with in the continuous machine when extruding lead alloys containing antimony, tellurium, or other alloying metals. Small quantities of the alloying metals deposit in the screw pitches after a short time of operation which cause an irregular flow of the lead through the die and render the sheathing useless.

A continuous hydraulic aluminum cable sheathing press the design of which has only recently been published, is shown in Fig. 53. With this press which is fed with solid billets, it is not so important to save the charging- or waiting time as this is of minor importance as compared with the setting period of lead. It is much more important, however, to obviate the stop marks – or to take measures which prevent the cable insulation from charring and eliminate a variation in the wall thickness while the die-block is released. The "breathing" of the die-block causes the mandrel to shift axially which often amounts to as much as 1 mm. Thus the relative position of the tapered mandrel in the die changes by this value when variations in the wall thickness of the sheath up to 0.1 mm will result.

The press is equipped with a top- and bottom hydraulic cylinder a and b being positively connected to each other by columns c. The double-acting main pistons d and e move the crossheads f and g. The container h is stationary and mounted in the container-holder i which in turn is fixed to the columns c by nuts j. The top movable crosshead f carries the ram k. The die-block l with hollow ram m rests on the bottom cross-head. The hollow ram m moves into a filling chamber n being connected to the extrusion chamber p through a ball check valve.

Ram k having been withdrawn and an aluminum billet having been fed into the extrusion chamber p, the ram is lowered and the extrusion cycle initiated. The aluminum passes through the valve port past the valve o into the filling chamber n and from there through the central bore of the hollow ram into the die-block l from which it emerges in the form of the sheath. While the aluminum is passed into the filling chamber, the die-block l is forced down. During this operation the movable crosshead takes its lowest position by coming to rest on the bottom column nuts. Shortly before the ram completes its stroke, the cylinder b is connected also to the pressure piping and the die-block l is moved up by the piston e. Now the hollow ram m forces the aluminum out of the filling chamber into the die-block l. Now the ram k is withdrawn slowly when valve o moves against the valve port thus closing the filling chamber against the extrusion chamber. The ram k is then withdrawn quickly to allow for another billet to be charged while the pressure in the filling chamber is maintained and the extrusion cycle continues. The press therefore works continuously. Having charged the new billet, the extrusion cycle is repeated, the valve is forced open again, the filling chamber n is refilled and the die-block l is brought back into its lowest position.

Chapter III

PRESSES FOR THE EXTRUSION OF PROFILES, WIRE AND TUBES OF HEAVY- AND LIGHT-METALS

(See Fig. 54 and 55)

The many successes which had in the course of time been achieved in the extrusion of lead pipe and wire, naturally directed attention to the possibility of employing extrusion to other metals, such as copper alloys in particular. GEORG ALEXANDER DICK, of Offenbach am Main, has been

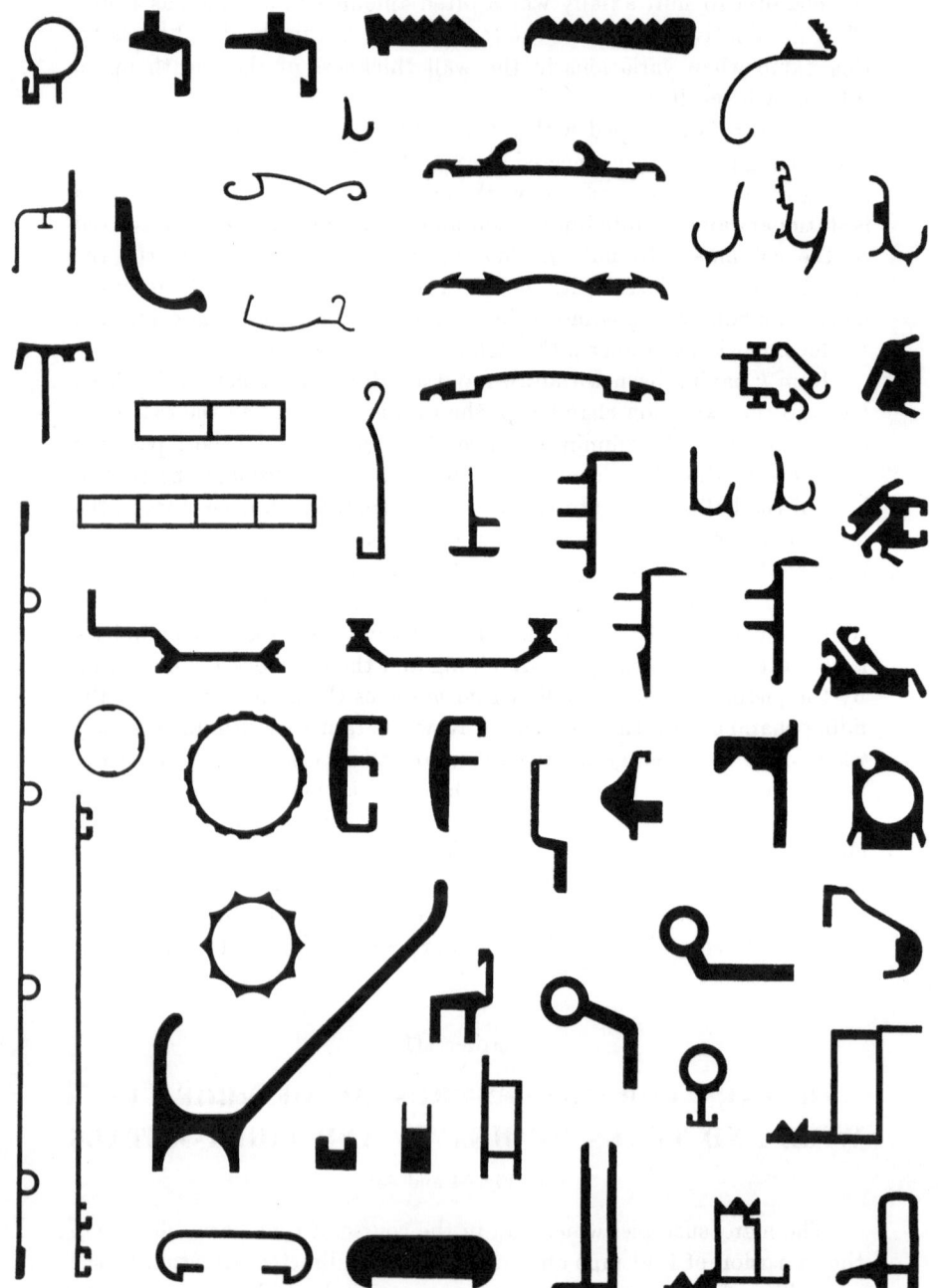

Fig. 54. Selection of solid and hollow sections, $1/3$ natural size (Osnabrücker Kupfer- und Drahtwerk)

Fig. 55. Light metal sections on runout and conveyor bed of a 12,000-ton tube and rod extrusion press
(Courtesy of Dow Chemical Co., Madison, USA)

one of the first trying to tackle this problem. In 1894 the first extrusion press[1]) was constructed according to his patent and design. It was capable of dealing with the following alloys:

	I	II	III	IV
Cu	58	60	85	90
Zn	40	40	10	–
Fe	2	–	–	–
Al	–	–	5	10

[1]) PEARSON, C. E.: The Extrusion of Metals. London: Chapman & Hall 1953.

This press was of the horizontal four-column type, fitted with a container out of which a cylindrical billet was forced through a die being locked in the counterplaten, by a ram being attached to the plunger. This extrusion method has been maintained in principle up to the present time and is generally known under the name of its inventor as *Dick Extrusion Method.*

In the early days considerable difficulty was met with the tools, owing to the inadequate material available for container, ram and die.

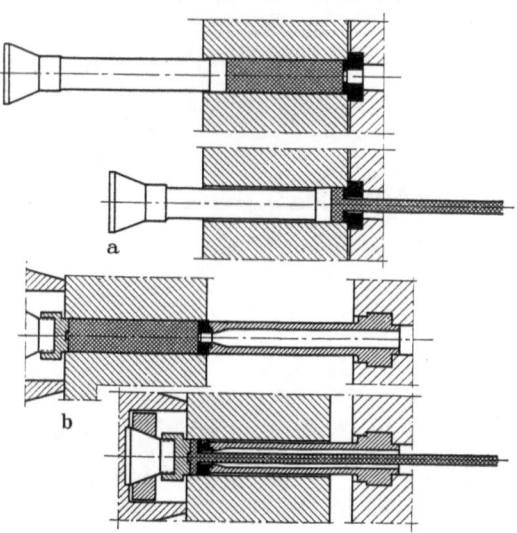

Fig. 56a and b, a) Direct extrusion (DICK's method),
b) inverted Extrusion

Any further succes was therefore dependent upon the production of high-quality steels suitable for high-temperature service. Such steels now meet highest demands.

The early extrusion presses were mainly used for the production of round rod and other solid sections. Tubes were manufactured by attaching a mandrel to the extrusion ram and forcing a billet which had been provided with a central bore, through an annulus formed by mandrel and die. The billet was not pierced in the container as the material available for the mandrels could not withstand the high stresses occurring. This method of extruding tubes was of course imperfect. Steps were, however, taken to improve machinery and materials which led finally to the combined tube and rod extrusion press in which the solid billet is pierced by the mandrel prior to extrusion. Simultaneously there were developed vertical presses for the extrusion of small diameter tubes.

Combined tube and rod extrusion presses are sometimes designed so as to be suitable both for direct and inverted extrusion. In the direct or DICK extrusion method, as illustrated in Fig. 56a, the billet lies between extrusion ram and die. Consequently, frictional resistance has to be overcome on the wall of the container which has its highest value at the beginning of the extrusion cycle and decreases in accordance with the reduction in length of the billet. When extruding by the inverted method, as shown in Fig. 56 b, the die is in front of the hollow extrusion ram while the opposite opening of the container is closed by a plate. During extrusion the resistance to deformation remains therefore unchanged and frictional losses on the container wall are avoided. Notwithstanding these advantages, the direct method has found widest application because in this case the size of the extrusion is not limited by the relatively small cross-section of the hollow ram.

a) Horizontal Presses for Direct Extrusion

Fig. 57 to 59 show the design of the first extrusion presses built in Germany about 50 to 60 years ago, which found a wide application even abroad. Some of these continue to be in operation even at the present time. In the press, shown in Fig. 57, a double-acting piston b made of chill casting, with extrusion ram c moves in a bronze-bushed, cast steel, hollow cylinder a. The ram having to be exchanged frequently and being made of costly alloy steel, it is provided – for reasons of economy – with a thrust piece. Sealing of the front part of the piston is ensured by a flange which holds a packing collar and is further provided with a connection for the return water. The cylinder bottom d takes the form of a crosshead and is connected to the counterplaten f by four columns e. The container-holder g is arranged in front of the counterplaten. It is guided on the columns and can perform a short longitudinal motion. For this purpose a double-acting piston h in the cylinder bottom pushes on a crosshead i which pulls the container-holder by two lateral rods j. The shifting stroke is limited by split column sleeves which are supported on the lugs of the cylinder a. The two piston faces are controlled alternately, i. e. during the extrusion stroke the container is forced against the die k in the counterplaten when sealing between container and die-holder is ensured by a cone; during the return stroke the container is stripped off the die thus allowing for the transverse slide l to be shifted. This transverse slide serves as abutment for the die-holder; it has a horseshoe-shaped aperture and is moved by hand. After unlocking the die-holder is ejected together with the dummy block and discard and pulled out of the platen. The discard is then separated on the table plate which forms an integrally cast part of the platen, using a hammer and a chisel.

When extruding cast billets, of which only the faces are machined, a "shell" forms in the container, the thickness of which is about 2 to 3 mm depending on the billet diameter. It contains oxides, dross and other casting impurities and eliminates the need for machining the billets. If this shell were to be pulled out of the container after the extrusion by means of the ram, a very high pullback power would be required. Therefore a loose dummy block, whose diameter is slightly larger than that

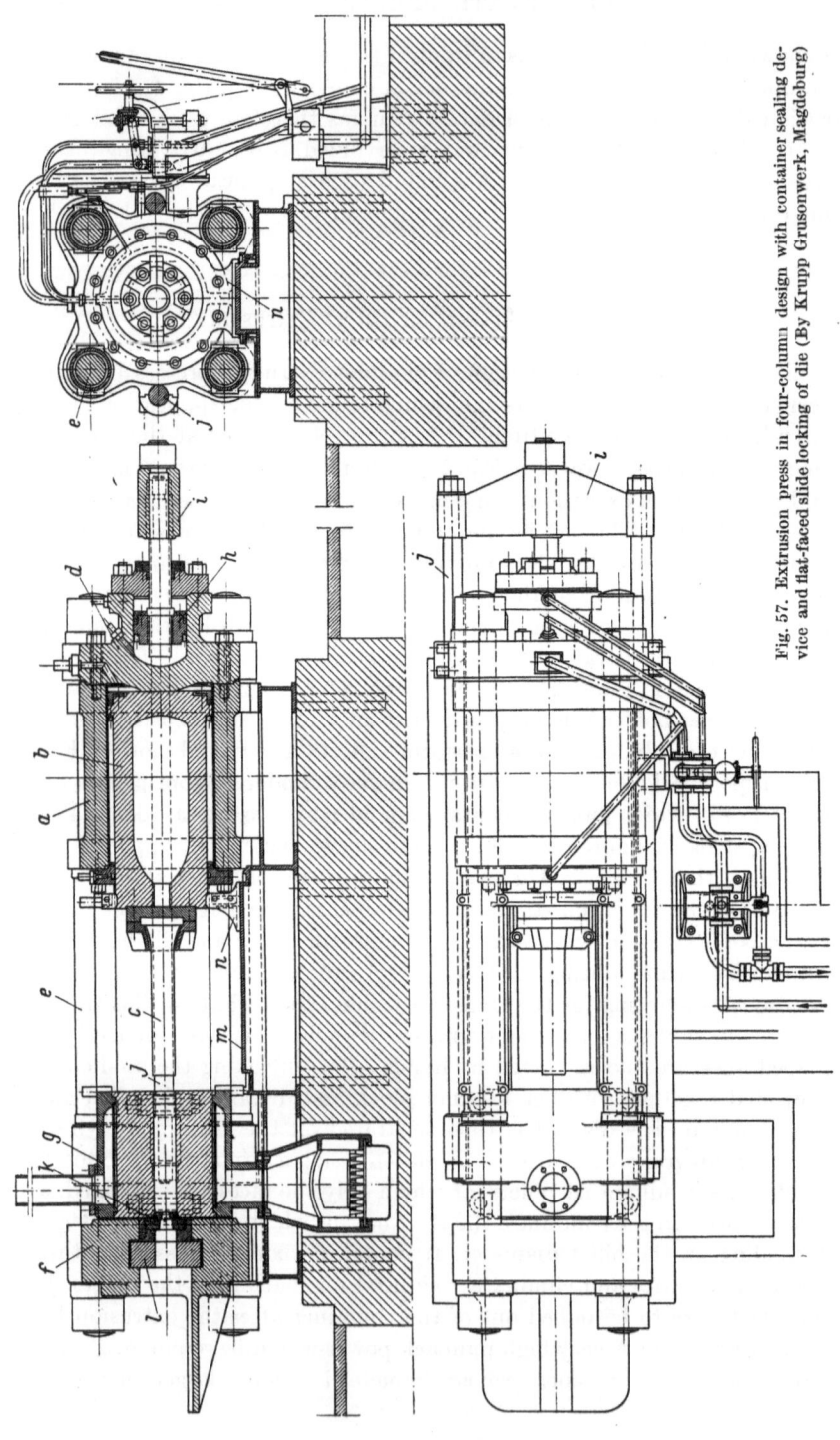

Fig. 57. Extrusion press in four-column design with container sealing device and flat-faced slide locking of die (By Krupp Grusonwerk, Magdeburg)

of the ram, is placed in front of the extrusion ram. After each extrusion this dummy is ejected together with the discard. For the purpose of expulsing the shell from the container, an additional idle stroke is performed after the ram has been retracted, when a close-fitting ejecting or clearing disk is placed in front of the extrusion ram and driven through the container bore.

When charging the press, the billet is inserted in the container from behind the press through the counterplaten. Then the die-holder is put

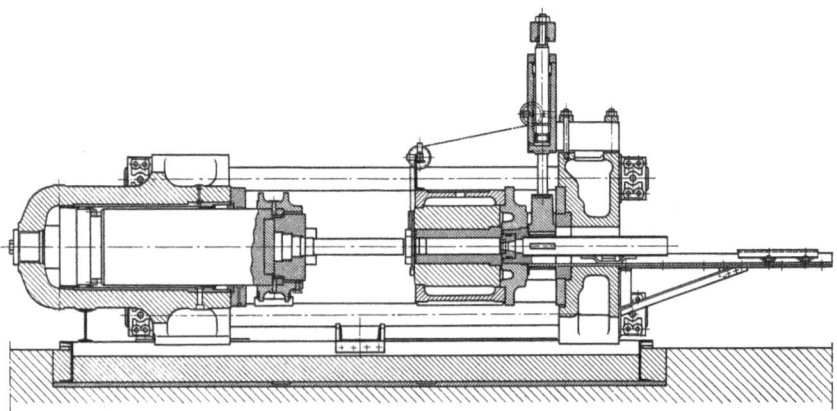

Fig. 58. Extrusion press in four-column design with fixed container and wedge-type locking device for the die

in place and locked by means of the transverse slide. The container is heated by a coal or coke burning furnace disposed underneath the container-holder. The fuel gases are circulated along either side of the container and passed into an exhaust pipe being attached to the container-holder.

The press rests on a base frame which is grouted with the foundation. It has a guide m for the plunger crosshead n which prevents a one-sided wearing of the sealing due to the weight of the piston.

The rod extrusion press is operated by a single-acting three-plunger pressure pump (see p. 204) designed for ram speeds of up to 15 mm/sec and an operating pressure of 400 atmospheres. The pressure water coming from the pump is fed into the different cylinder chambers via a hand-wheel operated control. A by-pass slide valve, operated by a hand lever, by means of which a connection between pressure and return piping is established, is arranged in front of the control. By means of this slide valve the control is relieved prior to actuating it and the press may be stopped in any desired position, when the pump will idle.

In the press, as shown in Fig. 58 and 59, the container is rigidly connected to the counterplaten by means of a jacket. The die-holder is set in a die-carrier which forces the former against the container liner by means of a horseshoe-shaped wedge. The wedge is suspended on a movable crosshead by two lateral rods and is lowered and raised by a double-acting piston. A stroke limiting slide, the movement of which is synchronized to that of the wedge, moves via a cable drive attached to the front end of the container. This slide ensures that the billet is always

Fig. 59. Shop photo of extrusion press, as per Fig. 58 (By Hydraulik, Duisburg)

extruded to a preselected thickness of the discard. With the die-carrier traveled out and the stroke limiting slide in its upper position, the discard may be ejected. Instead of the horseshoe-shaped wedge, two separately driven wedges, as per Fig. 60, were used in the early days. However, this design was not very successful and therefore abandoned soon.

The weight of the container and its jacket rests on the two bottom columns, on which the plunger crosshead slides, too. The ram may be adjusted in the plunger crosshead by means of setscrews so as to be able to compensate for any off-center position.

A drawback of double-acting pistons, as illustrated in Fig. 57 and 58, is that changing of sealings entails lengthy stoppages; also expensive packing collars, which have to be manufactured in molds, must be employed. Later designs of extrusion presses were improved by the application of single-acting plunger pistons which are sealed by stuffing boxes. Leakages are easily detected from the outside and quickly eliminated by simply retightening the gland. Packing material used comprises soft packings or vulcanized, relatively inexpensive sealing rings.

The improved cylinder design is shown in Fig. 61. The pullback cylinder is incorporated in the bottom of the main cylinder and cast integrally with the latter. For casting reasons it is provided with a full-length bore which in turn is closed by a plug. The pullback plunger *a* acts on a crosshead *c* being guided on the base frame *b*; the crosshead engages on the plunger crosshead *d* through two lateral rods. It is also possible to separate the pullback cylinder from the main cylinder and either insert the former into the latter or attach the former to the bottom of the latter. Other types of extrusion presses are equipped with two lateral pullback cylinders so as to reduce the overall length of the press; in this case, the pullback cylinders are provided either with stepped plungers with two stuffing boxes, as shown in Fig. 85, or with plungers with a single sealing and a reversing linkage, as per Fig. 67 and 77.

Charging the container by traveling the billet in on a small carriage, as shown in Fig. 58, as well as shifting the die-holder and separating the discard from the die using hammer and chisel, constitute very hard physical

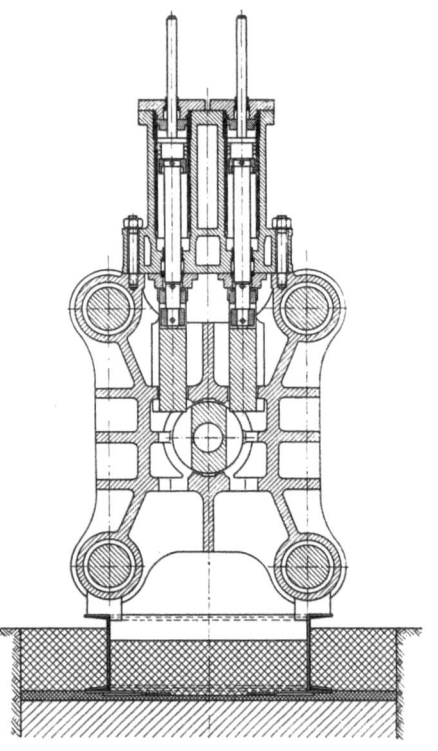

Fig. 60. Hydraulic drive for double-wedge locking of die

work. Soon the method of charging the press was changed in that the billet was no longer brought in from the rear, but from the side of the press, while the other motions were mechanized.

Fig. 62 and 63 show a rack and pinion drive for the die-holder. The die-carrier moves on two rollers running on rails which are fixed in the counterplaten and on the run-out table. The extension piece of the die-carrier consists of a tube, the top of which is cut open; the end of this tube is provided with a pinion shaft which is rotated by means of a hand-wheel. The racks of the double-sided drive are fixed on the run-out table.

In the course of time the hydraulic drive for the die-carrier was generally adopted. Fig. 64 shows an example of this in connection with a shears for the separation of the discard which is arranged immediately

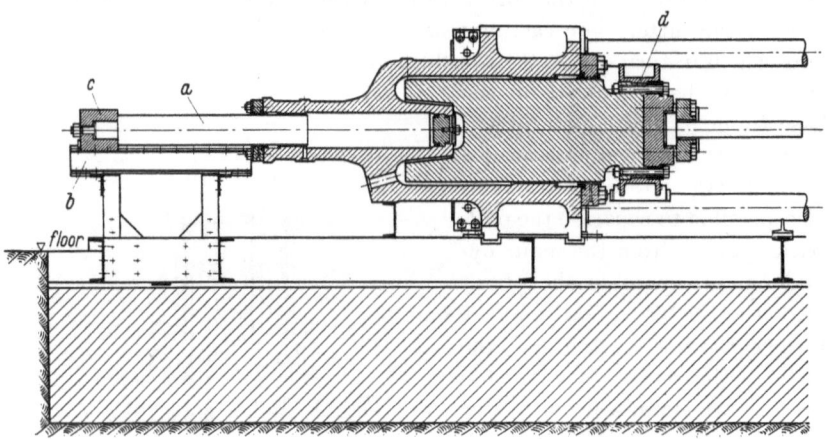

Fig. 61. Cylinder end of extrusion press with main- and pullback plungers being sealed by stuffing boxes

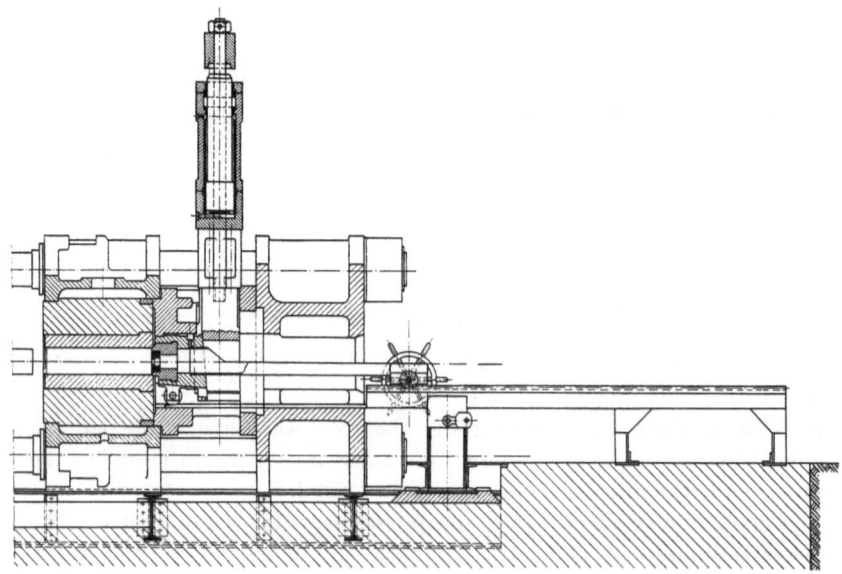

Fig. 62. Delivery side of extrusion press with rack and pinion drive for moving of die-carrier

behind the counterplaten. The run-out tube of the die-carrier is connected to a movable cylinder-block a which is securely guided on a long bed b. The plungers c and d are stationary and arranged one above the other. The run-out table *e* consists of a channel iron and is the continuation of the run-out tube; it is supported at several spots on the cylinder-block and its rear part is led over a roller. The cylinder-block has two stops *f* and *g* on the bed; the first stop fixes the position of the die-carrier in the shears and the second serves to set the end position in which the die is changed or inspected. When the die-carrier is traveled out, it first

Fig. 63. 2,000-ton extrusion press, as per Fig. 62, built in 1916. Courtesy of Berg, Werdohl
(By Hydraulik, Duisburg)

moves over the first stop. It is when the die-carrier is traveled in that its travel is limited in the shearing position so that it can never advance too far. The stop is retracted by actuating a foot pedal *h*.

The shears, as illustrated in Fig. 65, comprises of a yoke with two horizontal cylinders. In order to ensure a compact design and to eliminate a pullback linkage as well as front stuffing boxes which would be exposed to heat, the plungers are sealed at their ends by means of a collar; the pullback cylinders are incorporated in the sides of the plungers and secured to the shear cylinders. The pullback plungers, also sealed by collars, fit loosely in their cylinders; leakages are therefore detected on the outside. The two shearing blades cut the discard in the die plane.

In many cases it is more opportune to employ a saw instead of the shear; this saw is arranged either on a slide at the side of the bed or, as illustrated in Fig. 66, behind the counterplaten on a swinging lever. In general the shear is employed more frequently, because cutting is quicker

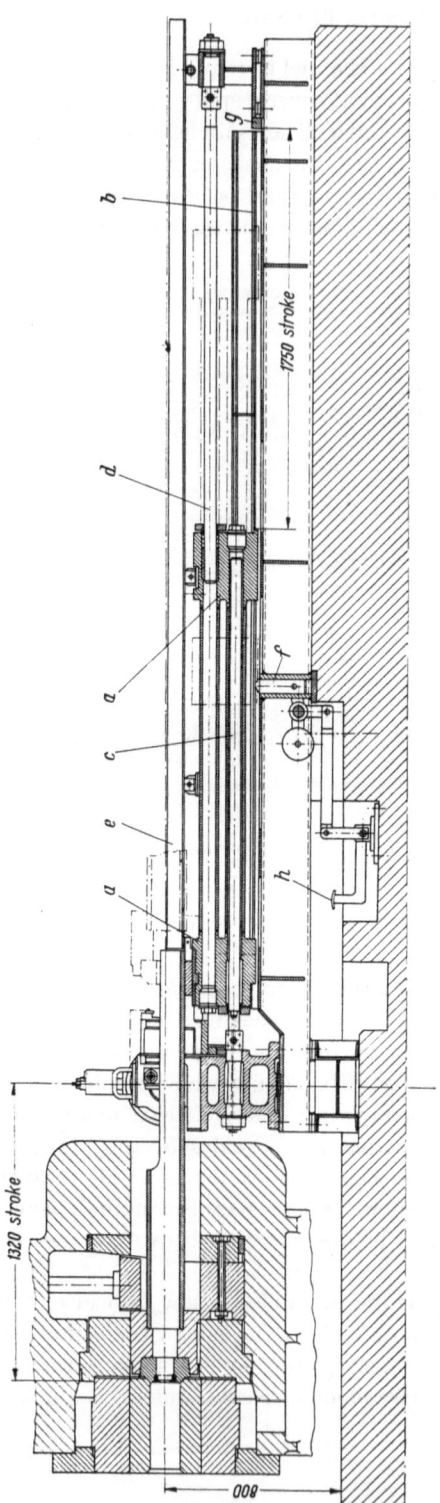

a, *b*, *c*, *d*, *e*, *a*, *h*, *f*, *g*

1750 stroke

1320 stroke

008

Fig. 64. Hydraulic die-carrier shifting device

than sawing. Moreover, many alloys cannot be sawn; they clog the teeth thus impairing badly the sawing capacity.

For the purpose of sawing, first of all a clearance between discard and die, into which the saw blade is to be inserted, must be established after the die-carrier has been traveled out. This is done with the help of a clamping device consisting of two levers *a* with jaws *b* which hold the discard when pressure water is fed into the small hydraulic cylinder *c*; when the die-carrier moves on, the extrusion is slightly pulled out of the die.

Hollow sections and tubes are most frequently sawn as they tend to bulge when being sheared, resulting thus in a loss which does not occur when sawing. Therefore, heavy presses are often arranged for either method of separation by equipping them with a combined shear and saw unit which is mounted on a common structure.

With the saw illustrated in Fig. 66, an electric motor *d* being mounted on the counterplaten, operates the saw blade *g* which is accommodated in a protecting cover and set in a swinging lever *f*, via V-belt pulleys *e*.

The point of rotation of this swinging lever is in the motor axle; it is pivoted by a linkage h which is operated from either side of the counter-platen by hand-levers i.

Fig. 67 illustrates an extrusion press of the "front loading" type, which may also be used for piercing of billets when extruding tubes; due

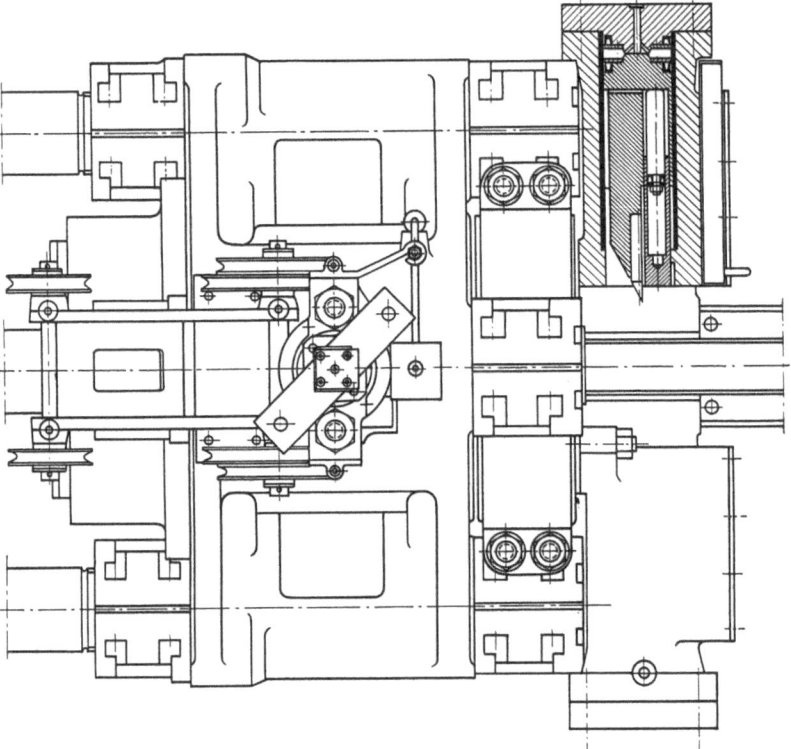

Fig. 65. Hydraulic shears on counterplaten with two moving blades to shear discard from the die

to the high piercing- and working speeds required it is no longer directly driven by a pressure water pump, but operated from a pressure water accumulator.

When extruding tubes, a piercing mandrel which is screwed in the ram, is used. Due to the projecting mandrel the extrusion billets may only be half as long as those used for rod extrusion which, on the other hand, increases the life of the mandrel.

The plan view shows two piercer cylinders on either side of the main cylinder; during the piercing operation the main cylinder is filled with low-pressure water of 2 to 3 atmospheres. This filling water is supplied

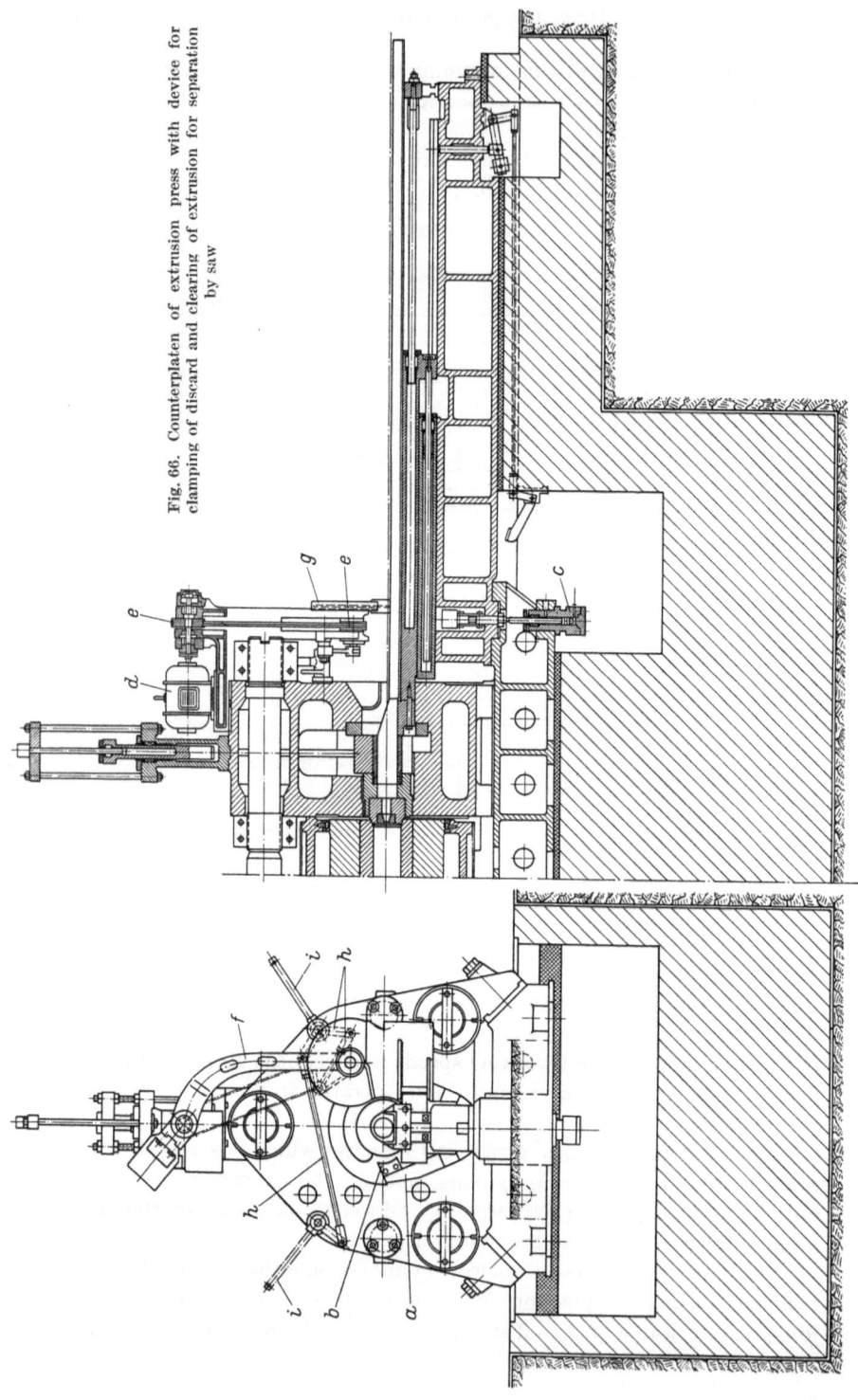

Fig. 66. Counterplaten of extrusion press with device for clamping of discard and clearing of extrusion for separation by saw

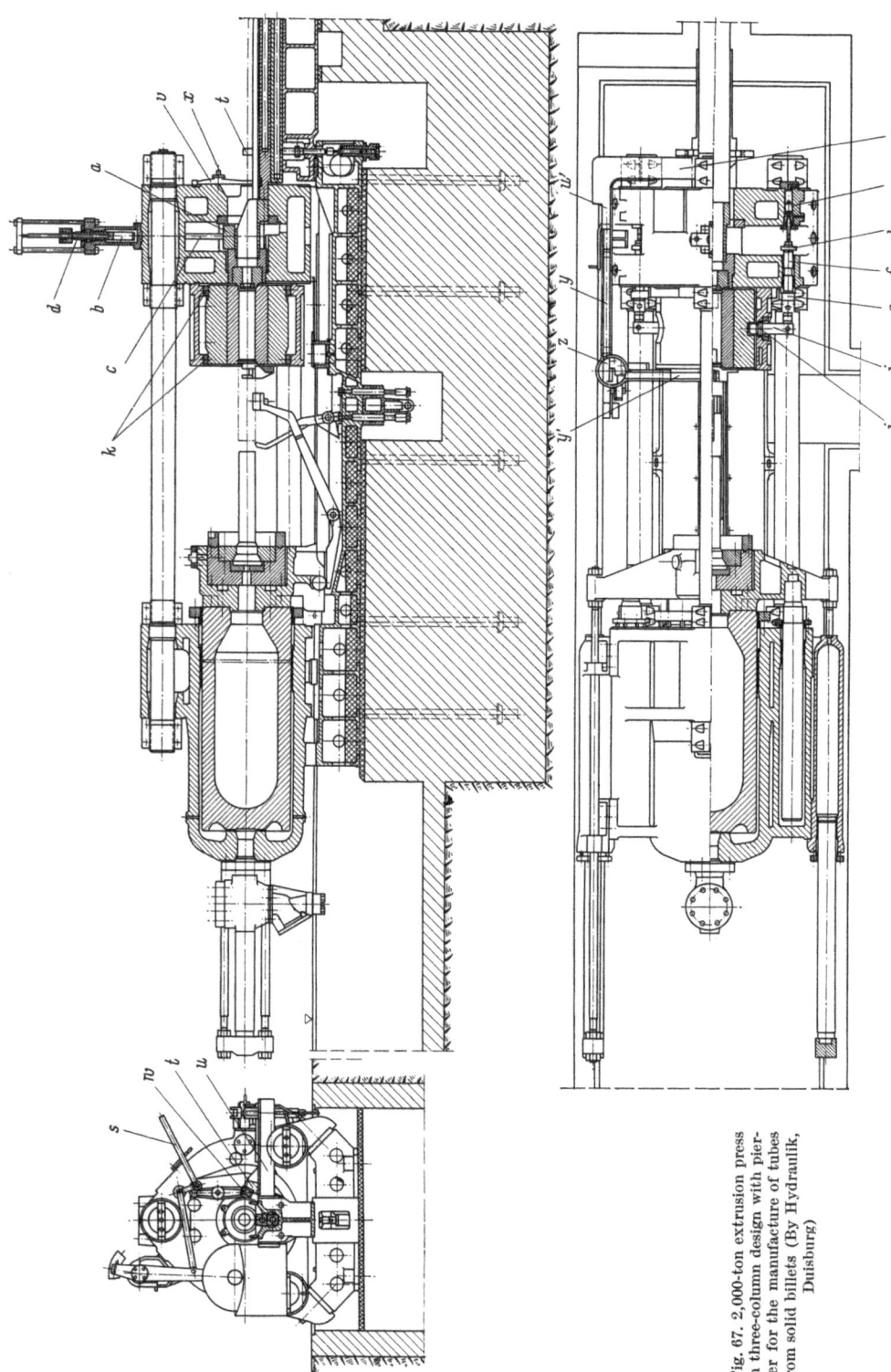

Fig. 67. 2,000-ton extrusion press in three-column design with piercer for the manufacture of tubes from solid billets (By Hydraulik, Duisburg)

by an air-vessel, arranged behind or beside the press, and enters the cylinder through a filler valve attached to its bottom.

The two pullback cylinders are arranged at the side of the two piercer cylinders. The pullback power is transmitted by the plungers through crossheads and rods to the ram crosshead which slides in adjustable guides on the base frame. The thrust carrying piece for the ram may be adjusted vertically and to either side over a given distance by means of setscrews, and locked in its various positions by paraxially arranged setscrews so that any misalignment of ram- and container axis may be corrected.

The counterplaten is connected to the cylinder-platen by three columns, the ends of which are kept prestressed by split nuts and lock washers (see p. 196).

The die-carrier is interlocked by a flat slide shown in Fig. 67, which, however, is no more moved manually, but raised and lowered hydraulically. The lifting piston b has a crosshead-shaped head to which the flat slide is fixed by two lateral rods c; it also forms the cylinder for the closing piston d. This type of locking the die-carrier offers the advantage – as compared with the wedge-type slide – that the bearing faces do not wear and that the concentric seating of the die is ensured to a higher degree due to the elimination of the wedge force components. This flat slide requires, however, a container sealing and stripping device consisting of a small double-acting piston e on either platen side, which operates a ram f with stops g and h and carrying a lever i which rotates around point j in the container-holder. For the purpose of guiding the container, the holder is provided – in the top and bottom central planes as well as at its sides – with keys k which lie in bronze-bushed grooves on the container and which permit thermal expansion in any direction.

The three-column construction was chosen to enable easy charging of the billet into the press. When loading the press, the billet is lifted from the furnace table with the help of a pair of tongs (see Fig. 63), being suspended from a monorail or I-beam by a chain and rollers, and placed on a charging cradle l in front of the container (see Fig. 68). This cradle rotates in m in the base frame n on which the ram crosshead is guided in its travel. Via a roller o the moving ram crosshead rotates the charging cradle. At the same time – via an arm – the ram crosshead shifts point p, being provided with a roller, of lever q which rotates in r thus pushing the billet into the container. Point r is cushioned and moves down when the lever q – after having pushed the billet into the container – is pushed down by the arm which continues its travel, in order to clear the container bore for the ram.

The tubes and rods are separated from the discard by a saw of the type already described; it is placed into sawing position (see Fig. 67) by means of hand-lever *s* after having first clamped the discard in the jaws *t* and established a clearance by further advancing the die-carrier. The saw cut having been performed, the discard is pulled on to table *u* with

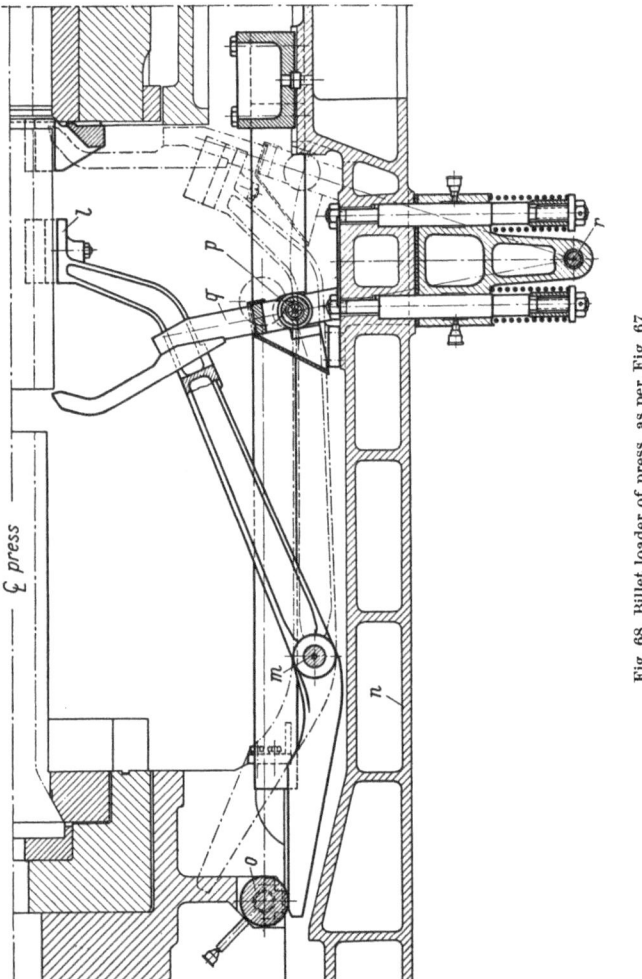

Fig. 68. Billet loader of press, as per Fig. 67

the help of a pair of tongs. For the purpose of removing the sawn-off rod out of the die, the lever *i* is pivoted about point *w* into the center of the press. The die-carrier is then traveled against plug *x* whose shape corresponds with the section of the rod. When using a shears severing the extrusion right on the die, ejection is in most cases not necessary. In this case the rods come off without having to apply any considerable amount of force; in many cases they are left in the die and pushed through by the subsequent extrusion.

Separation of the dummy from the "shell" is very hard physical work. It may be facilitated by a separator press placed at the side of the table, as shown in Fig. 69. This separator press consists of a housing a with a spindle b serving to adjust the dummy c in the "shell" d to the edge of the blade e. The blade is moved by a hydraulic piston f and slides

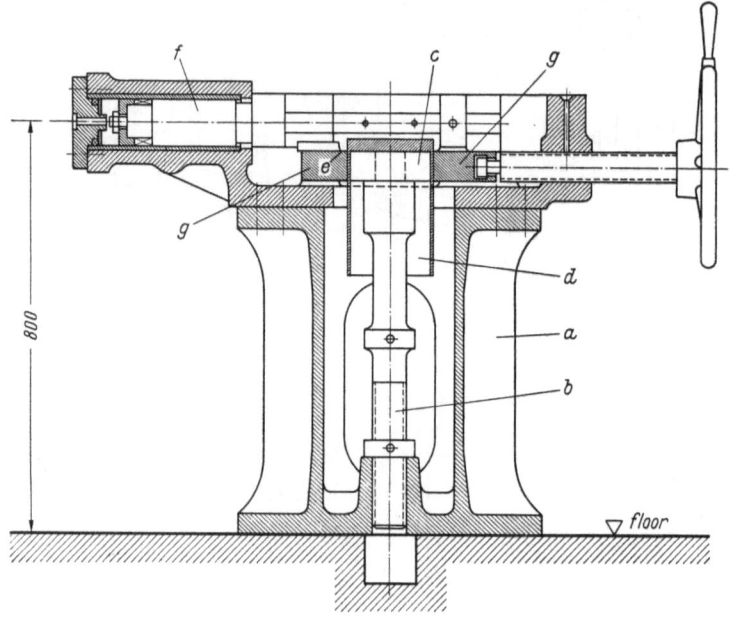

Fig. 69. Separator press for separation of discard and dummy-block

over the clamping jaws g which are opened and closed by spindle and hand-wheel.

The discard having been sheared off, the dummy is knocked out of the "shell" and placed upright on a small platform u' (see Fig. 67), from where it is lifted by actuating a hand-lever, whereupon it rolls off through an inclined chute y on to a rotary table z and after a $90°$ turn through the chute y' right in front of the container. The rotary table is moved by actuating a foot pedal. This chute y' is split to allow for the container to be shifted.

Conveying the billet from the furnace to the press in the manner described in the foregoing, is still connected with quite some effort on the part of the operator. That is why mechanical devices were developed in the course of time. Two examples are shown in Fig. 70 and 71. Having

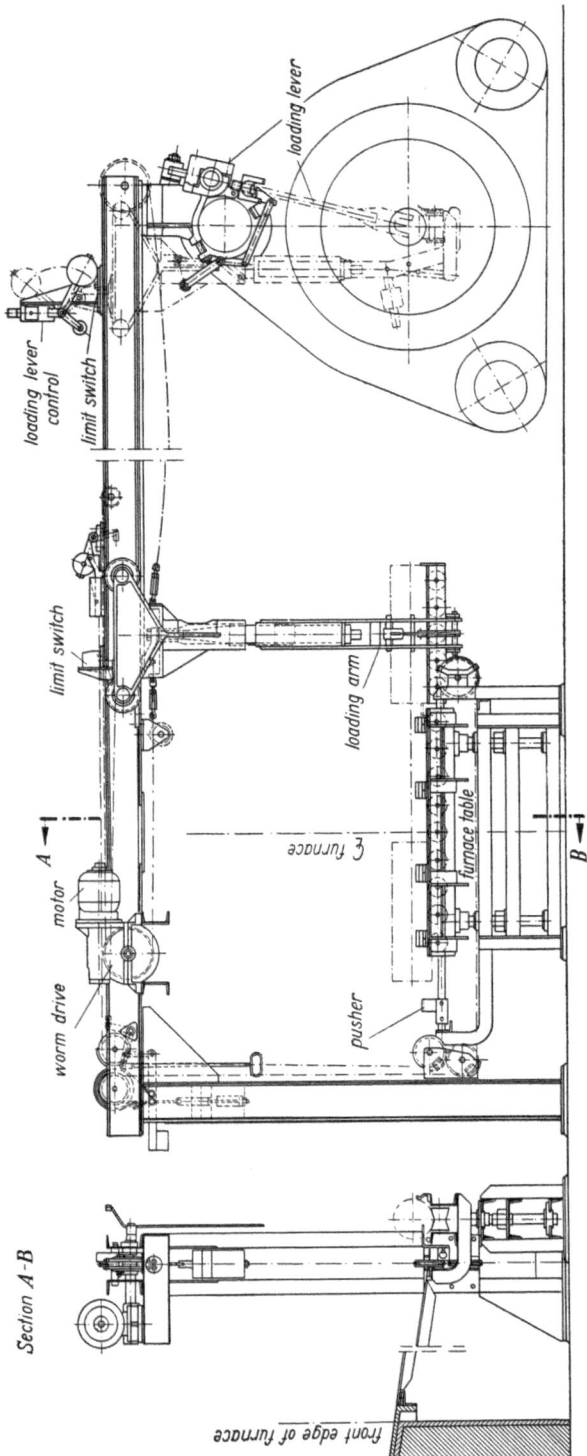

Fig. 70. Billet conveyor

once been initiated, the charging operation is performed fully automatically. From a pusher-type preheating furnace (see Fig. 70) the billet rolls on a small roller bed which can be adjusted vertically to suit the various billet diameters. The billet is then pushed by a stem on a loading arm

Fig. 71. Shop photo of billet conveyor (By Schloemann, Düsseldorf)

being provided with rollers. The stem having reached its end position, the loading arm starts moving, while the billet is turned by 90°. The loading arm stops in the center of the press. Then a lever pushes the billet into the container and the loading arm returns to its home position.

The runway consists of two steel channels which are supported on the top column of the press and a girder at the furnace. The loading arm and the stem are operated by a sprocket chain and a geared motor. The loading lever is operated by a hydraulic piston.

Fig. 72 shows the hydraulic circuit diagram of an extrusion press operated from a pressure water accumulator. The press has separate controls for the piercing- and main cylinders. When working without piercer, the control lever of the piercer control is set to *return*. When the lever of the main control is shifted to *advance*, the filling water flows from the air-vessel through the filler valve into the main cylinder and through a check valve into the two piercer cylinders and advances the plunger at no-load, while the water contained in the pullback cylinders flows back through the opened valve 2 of the main control. As soon as any higher resistance is encountered, the filler valve closes so that valve 3 has to be opened to advance the plunger further. A double-seat throttle valve, arranged behind this valve, is operated by another hand-lever and permits of adjusting the ram speed at random. When the ram (plunger) is to be returned, valves 1 and 4 are opened by means of the main control lever; thus the main cylinder is relieved, pressure water is fed to the pullback cylinders and the cylinder of the driving piston is connected to the pullback piping, whereby the driving piston opens the filler valve. During the return motion the water flows out of the pullback cylinders through the opened valve 2 of the piercer control back into the air-vessel. The sequence of the valve lifts is illustrated in the valve lift diagram.

When piercing billets, the lever of the main control is first shifted to *advance* and the idle stroke having been completed, the lever of the piercer control is shifted to *piercing* etc.

Simple two-valve controls suffice to effect the piston motions of the auxiliary equipment of the press – with the exception of the die-carrier shifting device – as the pullback ends of the pistons are suitably connected directly to the pressure water circuit (see p. 221). The sectional areas of the pistons of the shifting devices are of equal size and therefore require a four-valve control.

Fig. 73 illustrates the design of a heavy extrusion press of about 4,000-ton capacity, suitable for the manufacture of light-metal rods, wire and strips. It is in three-column construction and equipped with a wedge-type locking device for the die-carrier and a combined shear and saw to separate the discard. The shearing blade cuts downwards and the driving cylinders are mounted in the top part of a frame which is attached to the bed of the shifting device. The frame side which faces the counterplaten, is designed in the form of a guide for the hydraulically movable slide which carries the vertical driving motor for the saw blade.

The billet loading device is always designed to suit the respective conditions. In the foregoing design the billet comes from a roller-type preheating furnace and rolls over a roller table at right angles to the pressing direction on to a roller table which is located in the center of the press; this table is lifted by two lateral cylinders and at the same time turned by 90 °. For this purpose the center guide rod is provided with a steep thread which rotates the rod in a nut. In the stroke end position the ram pushes the billet into the container. Then the ram is slightly withdrawn to allow for the dummy to be placed before the billet. The dummy is conveyed through a chute from where it rolls on to a tray which is attached to a hydraulically operated lifting mechanism in front of the roller table.

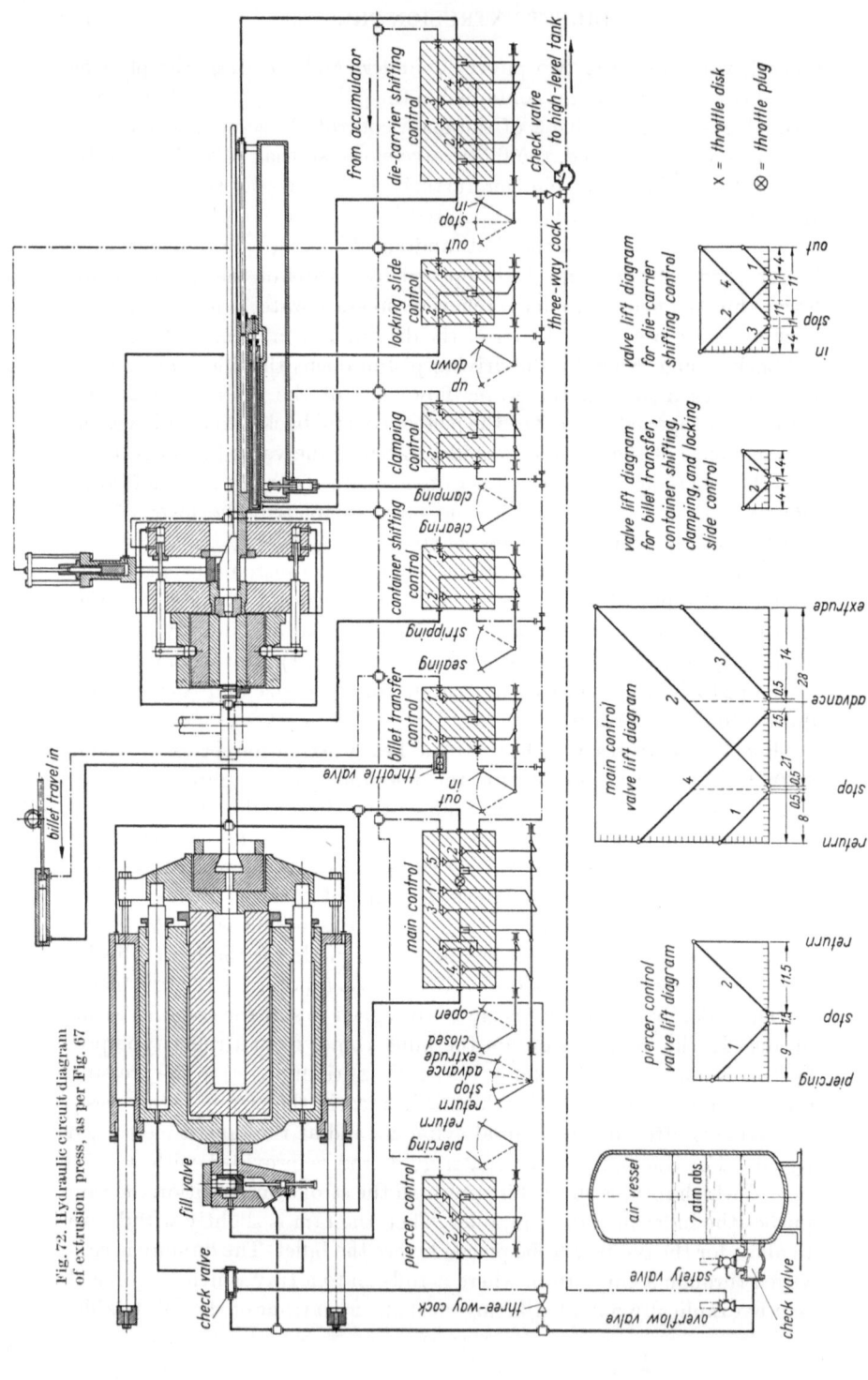

Fig. 72. Hydraulic circuit diagram of extrusion press, as per Fig. 67

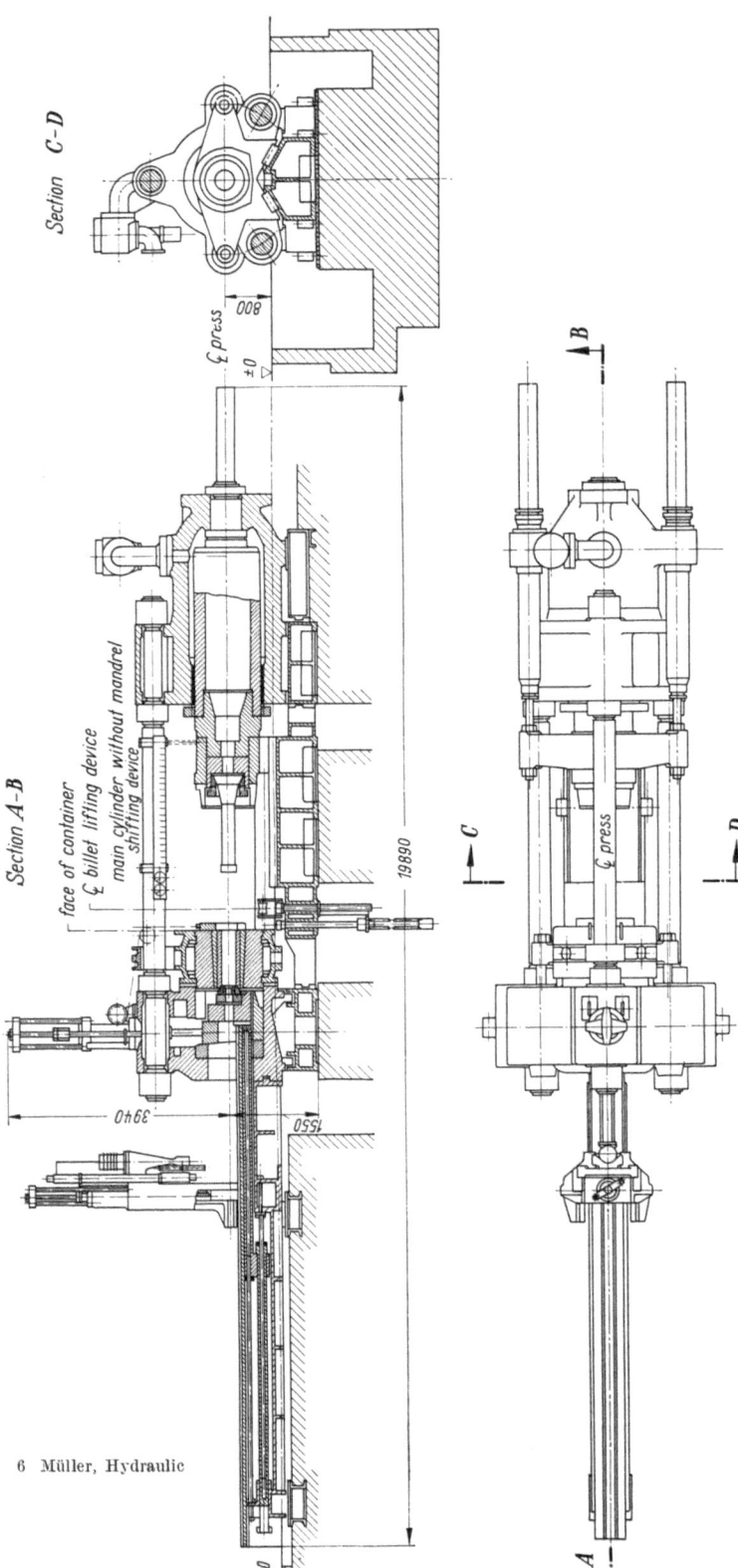

Section C-D

Section A-B

face of container
₵ billet lifting device
main cylinder without mandrel
shifting device

₵ press

800

±0

19890

3940

1550

±0

A

B

C

₵ press

D

6 Müller, Hydraulic

Fig. 73. 4,000-ton extrusion press in three-column design with die-carrier shifting device and combined shear and saw arranged on the bed of the shifting device (By Hydropress, New York)

If hollow sections and tubes are to be extruded on the press shown in
Fig. 73, shorter billets will have to be used, because due to the mandrel
fixed in the ram and the space required for loading the billet from below,
the stroke of the press is reduced by about one billet length. In order to
remedy this drawback, a mandrel shifting device may be incorporated
in the main plunger of the press, as illustrated in Fig. 74. This device is
of particular advantage when extruding tubes or hollow sections from
light-metal billets, into which a hole has been drilled or which are hollow-

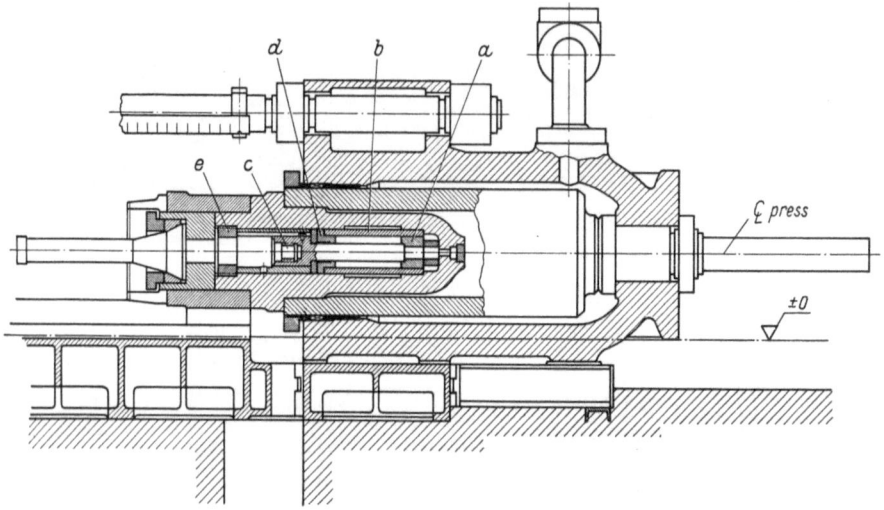

Fig. 74. Extrusion press, as per Fig. 73; however, with mandrel shifting device arranged
in the main plunger

cast, and with which no high resistance has to be overcome when advancing the mandrel into the die. This device consists of a double-acting
piston a, moving in a cylinder-bush b and having at its front end a tap-
hole c into which the mandrel-holder is screwed. The mandrel-holder is
guided in the ram and serves to hold the mandrels of various diameters.
Stuffing box d seals the piston-rod and all of the internals may be removed
out of the plunger after nut e has been released. The two cylinder connections are on the main plunger and the connection to the control unit
is established by long plunger pistons, so-called telescopic or articulated
pipes. Maintenance of such moving joints not being easy, preference is
in most cases given to a better, though more expensive design, as illustrated in Fig. 95, which offers another advantage in that the mandrel may
be arrested in a relative position in the die.

Latest development of extrusion presses is characterized by a marked
trend towards the adoption of fully or semi-automatic operation and by

an increase in production by cutting down the idle times, as well as by a trend towards the utilization of oil drive instead of water drive, thus leading to simpler and cheaper presses (see also p. 207). The idle times which may be reduced, are mainly limited to the times required for advancing and locking the die-carrier. However, the percentage of these times in the complete working cycle decreases with the size of the press increasing, so that marked advantages by a reduction of these times are only obtainable with extrusion presses up to about 2,500-ton capacity. With such presses the times required for traveling out the discard and cutting off the rod may be reduced by providing a lateral slide at the counterplaten instead of the axially moving die-carrier. This also entails a change of the shear or saw arrangement from the rear to the front side of the platen.

A typical example of a lateral slide is shown in Fig. 75; it is provided with two holes, of which one serves to accommodate the die and the die-holder. Shifting is performed by swinging out a lever, whose driving piston and cylinder are arranged on the counterplaten so as to protect them against heat. The stroke of the lateral slide is essentially shorter than that of the die-carrier and with the die traveled out, the shell may be ejected out of the container by the ram through the other hole. The traveled-out die may easily and quickly be inspected and lubricated; if it is to be exchanged, it is ejected from the die-holder with the help of a small ejector ram arranged on the counterplaten, whereas it had formerly to be detached by blows with a rod.

Fig. 76 shows the arrangement of a shear in front of the counter-platen. Prior to shearing the container is shifted on the base frame and the discard retained by the ram. Shearing having been completed, the rod is pulled out of the die using a pair of tongs; discard with dummy slide over a chute to a point beside the press.

If the dies have to be changed frequently, it is advantageous to use a turret die-holder equipped with two opposed stations, i. e. four holes, for the accomodation of the dies and the ejection of the shells. This arrangement permits simultaneously of one die being used for extruding and of the other die being cooled or changed. Also, it would be possible to arrange the two dies in a horizontal slide with the ejector hole in between; then, however, changing of dies would be rather complicated and would have to take place on either side of the press[1]). Installation of such rotary head is relatively simple in a three-column press in that one column, about which the rotary head would be pivoted, is arranged in the horizontal center plane. In this case, however, one has to put up with the disadvantage that the access to the press would be obstructed

[1]) GEIGER, W.: Entwicklungsstand hydraulischer Pressen. Z. Metallkde. Bd. 45 (1954) H. 1.

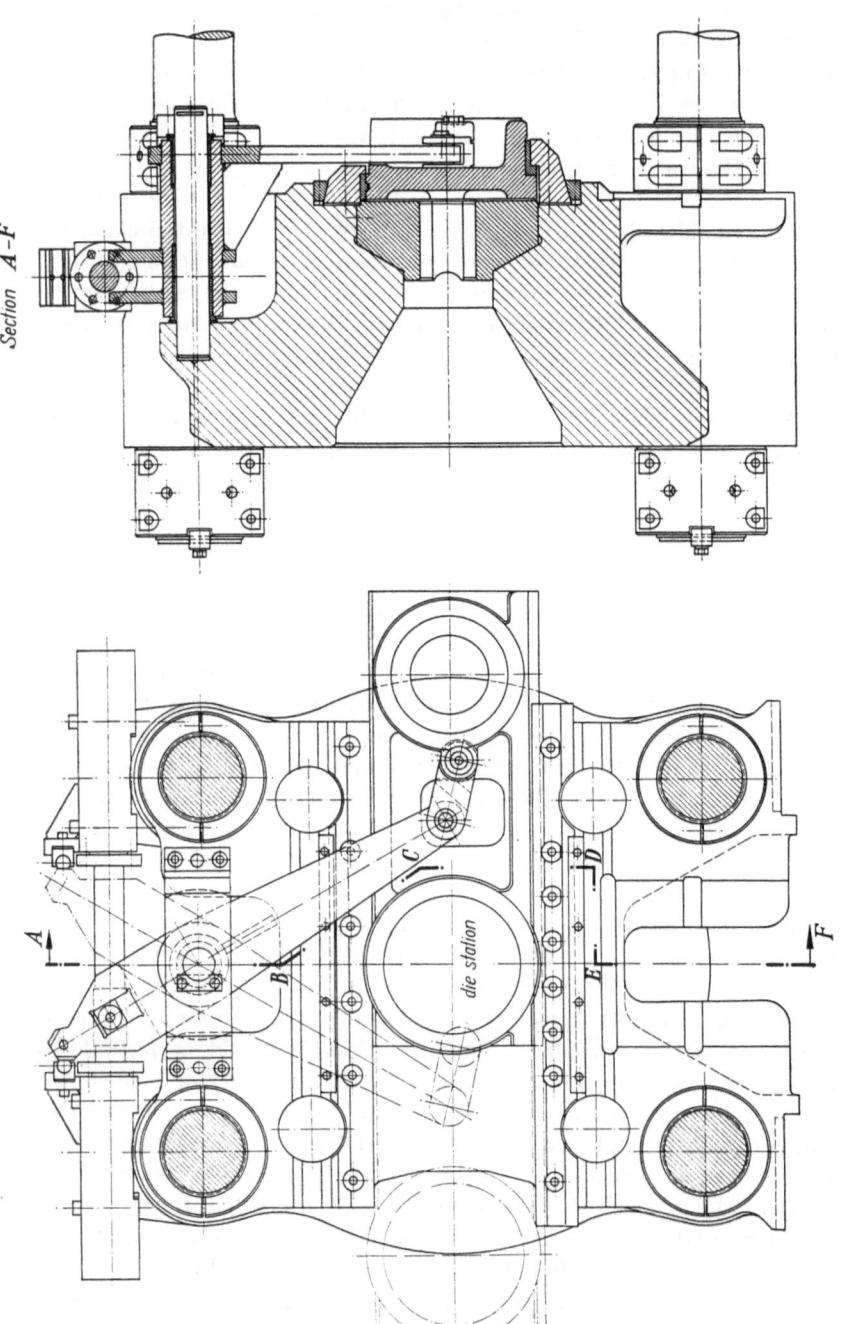

Section A-F

die station

Fig. 75. Lateral die slide on the counterplaten of a four-column extrusion press

and that billet- and dummy feeding may be effected from one side of the press only (see Fig. 150).

All of the extrusion presses described so far, are in column construction. However, presses of up to 1,500-ton capacity are also built in frame

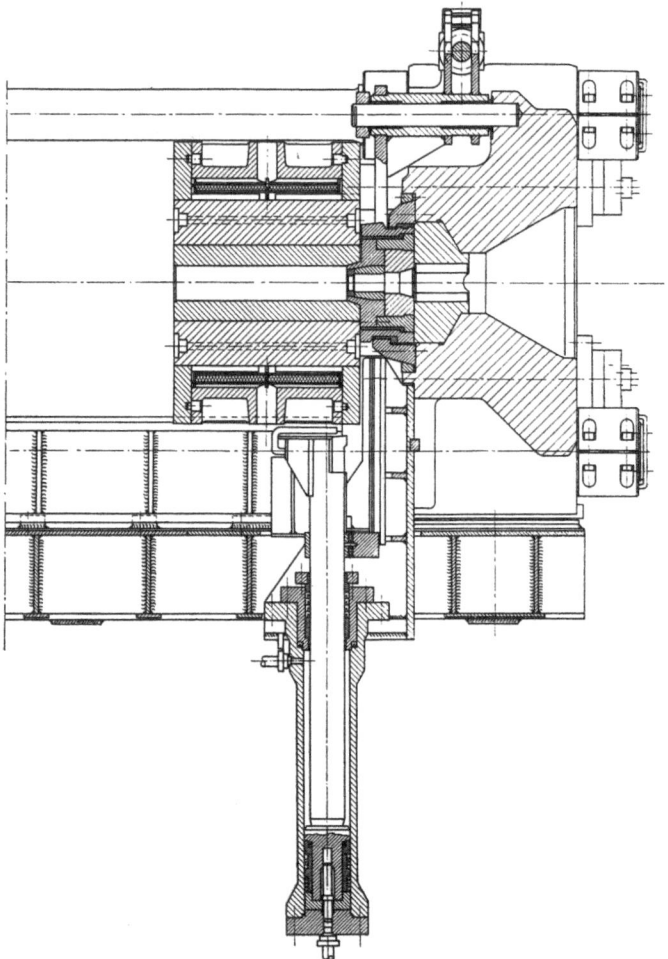

Fig. 76. Up-acting shear in front of the counterplaten

construction (see Fig. 77). This type offers the advantage that all of the axial bores can be drilled in tandem on the boring mill and are therefore in perfect concentric alignment. Thus an optimum of accuracy in machining is obtained, this being of particular importance in the production of tubes and hollow sections. This design is in many cases preferred for small oil-hydraulically operated presses, since the complete drive unit may be mounted on the frame which results in an enclosed, compact machine. Choice of column- or frame construction is furthermore a question of price.

Fig. 75 to 80 show a 500-ton press in frame construction. It is of the "front loading" type and its outstanding features are double-acting pis-

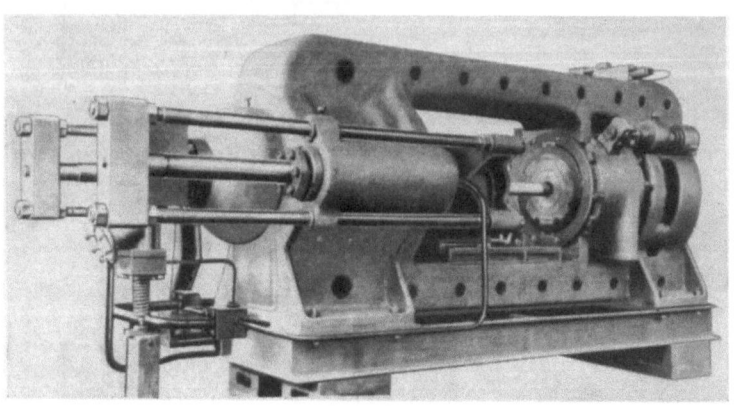

Fig. 77. Shop photo of a water-hydraulic 1,500-ton extrusion press in frame construction
(By Hydraulik, Duisburg)

tons in the main cylinder and in the die-carrier shifting device, being sealed by cast-iron piston rings. This type of piston sealing may only be used when the operating medium is oil (see p. 237). This has been one of the contributing factors for the wide application of oil-hydraulic presses in metalworks.

A frame-construction press equipped with a horizontal die slide is illustrated in Fig. 81. The slide has again two holes, one for the die and one for the ejection of the shell. With the slide traveled out fully, the die-holder is removed towards the rear and pushed on a trough, as shown in the diagram. The slide is shifted by a double-acting hydraulic piston with its appertaining cylinder being mounted under the slide on the guideway. This arrangement is, however, recommended only for the extrusion of metals at a temperature up to about 500 °C; with higher temperatures the arrangement shown in Fig. 75, is more suitable.

The shear which serves to sever the discard and the dummy from the extrusion, is mounted oblique on the base frame. At the side of the shearing blade there is arranged an arm which moves on a roller and pushes the discard with the dummy into a receptacle or on a table when the shearing blade is traveled out. The bolt in the second opening serves to secure the slide during the shearing operation.

Fig. 82 illustrates the hydraulic circuit diagram of an oil-hydraulic extrusion press operated by a variable-delivery pump (see p. 212). The control stand accommodates a hand-wheel connected linkage which sets the press to *advance* or *return*, as well as two rotary slide valves which serve to lock and shift the die-carrier with the die. The linkage acts on a servo-control (see p. 215), which is arranged directly at the pump; the servo-control is connected to a geared pump, also called pilot pressure pump.

During the idle stroke the oil flows from the filling tank through the filler valve into the cylinder and forces the main piston of the press forward at a pressure of about 5 atmospheres. The oil which is forced out of the pullback chamber of the cylinder, is fed back into the filling tank by the pump. The idle speed may therefore be adjusted at random by simply changing the pump stroke.

The pilot pressure is furthermore most suitably employed to perform the auxiliary motions when shifting and locking the die-carrier. It continuously delivers oil – against the

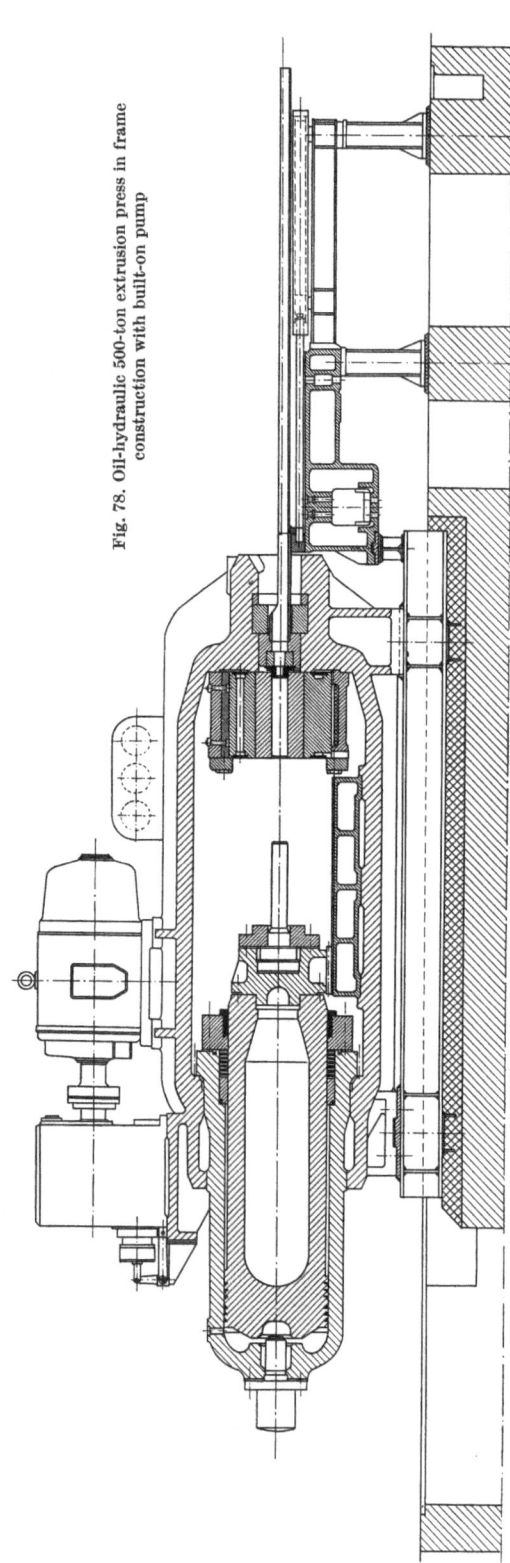

Fig. 78. Oil-hydraulic 500-ton extrusion press in frame construction with built-on pump

pressure of a spring-loaded by-pass valve – to the two rotary slide valves from where it is fed into the cylinders. Another geared pump – the so-

Fig. 79. Shop photo of oil-hydraulic extrusion press, as per Fig. 78 (By Hydraulik, Duisburg)

Fig. 80. Delivery end of press, as per Fig. 79

called circulating pump – is connected to the pilot pressure pump. This second geared pump is rated for a maximum pressure of about 10 atmospheres and pumps the complete leak oil – which comes from the main

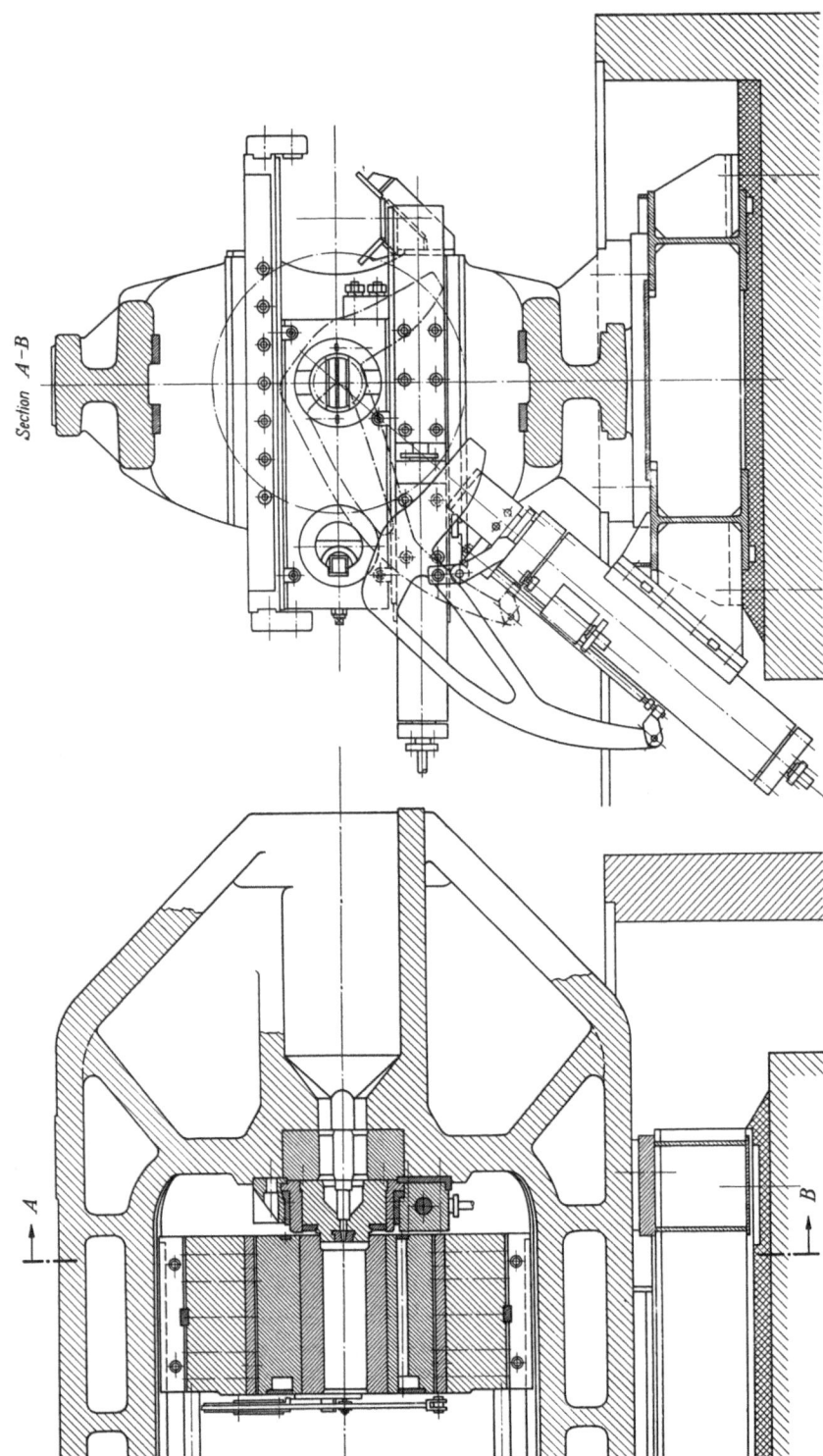

Section A–B

A

B

Fig. 81. Oil-hydraulic extrusion press in frame construction with lateral die slide and shear in inclined arrangement

pump, all of the cylinders and control units, and is collected in an open reservoir – back into the filling tank through an oil-cooler and filter.

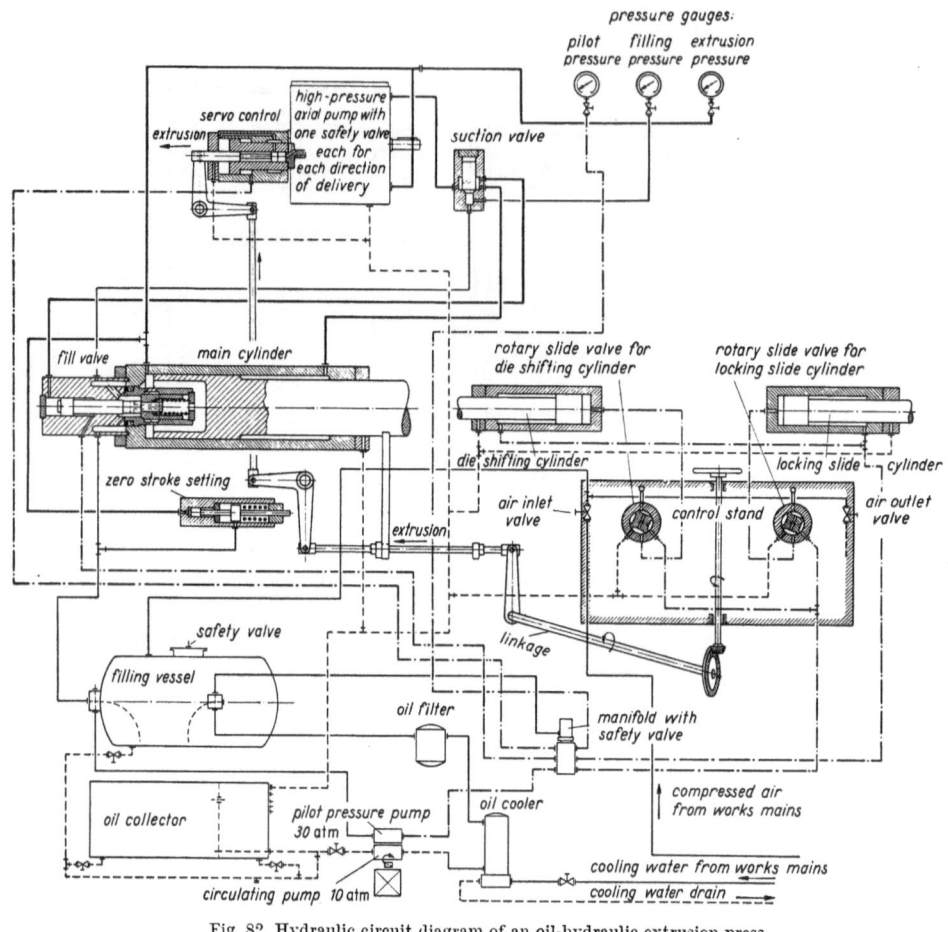

Fig. 82. Hydraulic circuit diagram of an oil-hydraulic extrusion press

| ——— | extrusion pressure | 200 atm | ——— | filling pressure | 5 atm |
| –·–·– | pilot pressure | 30 atm | – – – – | return oil | 0 atm |

b) Horizontal Presses for Direct and Indirect Extrusion

All of the extrusion presses dealt with so far, work on the direct method of extrusion, in which all of the billet particles move and, if billets of great length are extruded, an essential portion of the extrusion power is consumed by the skin friction in the container.

Fig. 83 and 84 show one of the earliest presses suitable for both direct and indirect extrusion. The billet is charged from the rear. The counterplaten is stationary and provided with the familiar device which locks the die-carrier by means of a wedge. It holds three columns which

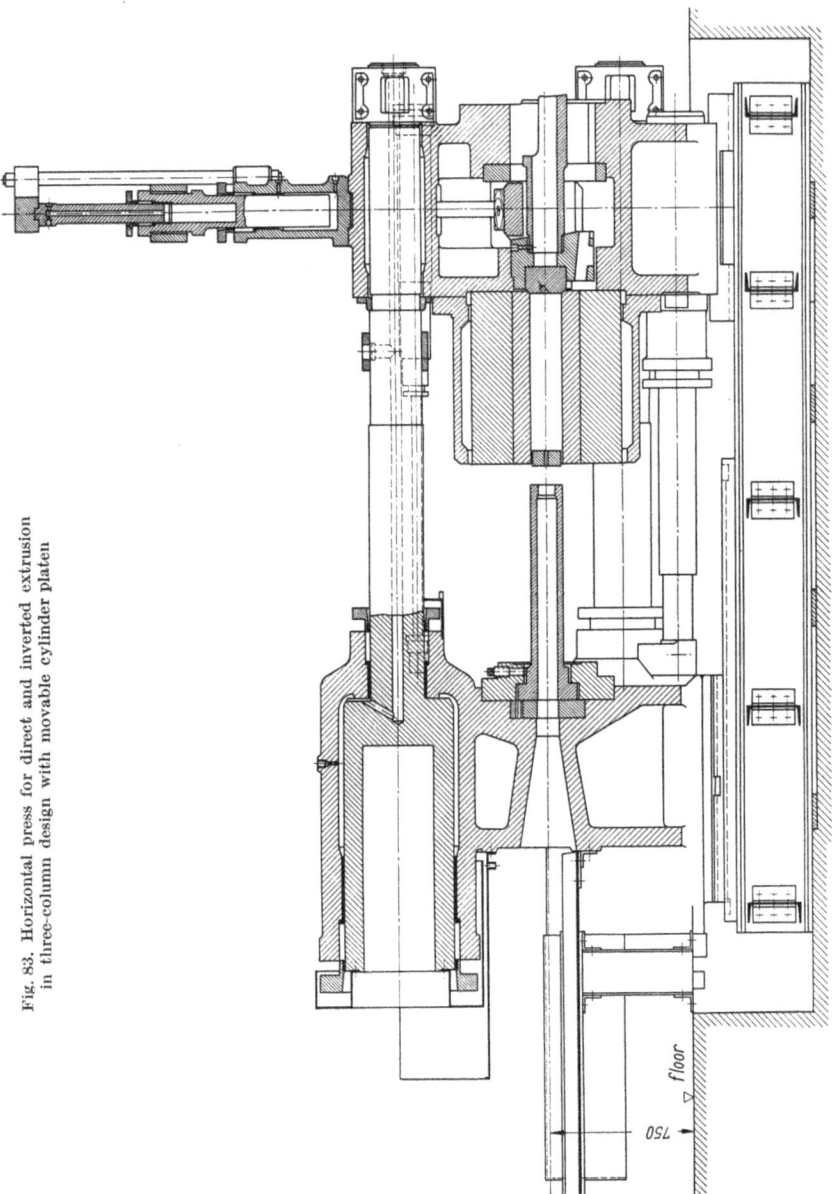

Fig. 83. Horizontal press for direct and inverted extrusion in three-column design with movable cylinder platen

at the same time serve as plungers for the moving cylinder-platen. The latter is guided in its travel on the base frame. The columns are inserted in the counterplaten with a little clearance and solidly supported in bronze bushes in each of the cylinders. They are further provided with a concentric bore for the pressure water supply. Leak water which may issue from the retightenable stuffing boxes, is collected in sheet metal pans. Two advance cylinders and two pullback cylinders are arranged in the counterplaten to effect the idle- and return strokes. Both main- and advance cylinders are equipped with differential pistons.

Fig. 84. Shop photo of press as per Fig. 83 (By Hydraulik, Duisburg)

The axis of the ram may be adjusted by means of setscrews. When extruding indirect a solid disk is mounted in the die-carrier instead of the die and the die-holder. This solid disk is provided with a small chamfer on which the die and the die-holder are caught when the die-carrier is traveled out.

Behind the counterplaten there is a die-carrier shifting device – not shown in the diagram – with the same type of runout table as used for direct extrusion, as well as a shear to cut off extrusion and discard.

The design which has been described in the foregoing, is only recommended for extrusion presses up to 500 tons capacity. With heavier

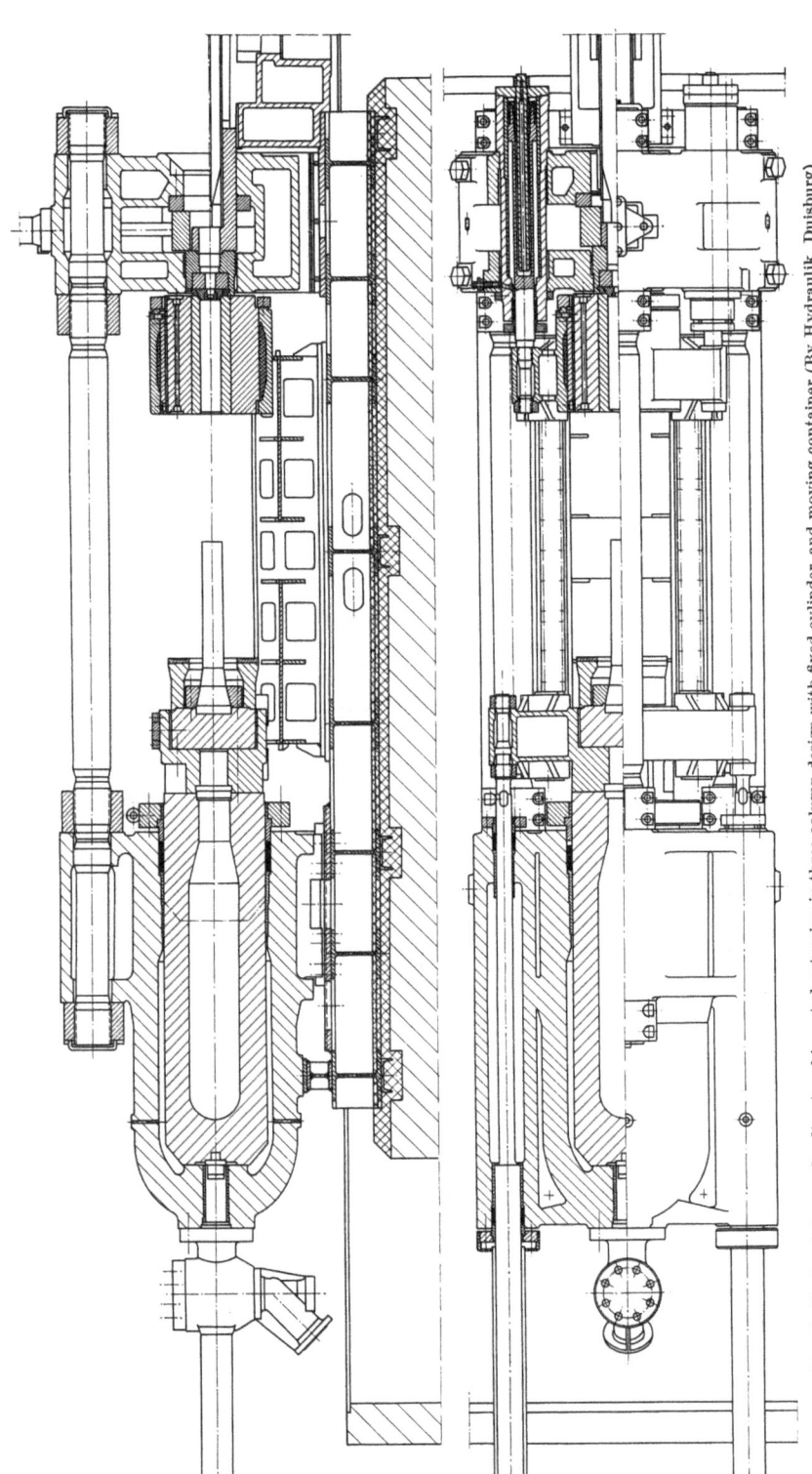

Fig. 85 . Horizontal press for direct and inverted extrusion in three-column design with fixed cylinder and moving container (By Hydraulik, Duisburg)

presses this type of runout table and the three-cylinder design have proved to be disadvantageous. The resistances in the cylinders vary almost
always and the difference in the extrusion pressures leads to a tipping
moment which causes heavy stresses on the guides and leakages in the
stuffing boxes. Heavy billets are furthermore not easily loaded in the
container from the rear of the press.

These drawbacks are eliminated in the design illustrated in Fig. 85,
which is, however, more expensive. This press is of the front loading type.
Fig. 85 shows the press when working on the direct method. The main
difference from the conventional design is that the container may be shifted
over a long distance on the guiding-bed of the plunger crosshead, thus
permitting of securing – for indirect extrusion – a ram instead of the
die on the die-carrier. This die-carrier is locked in the counterplaten by
a flat slide the design of which has already been described.

The container is shifted by two horizontally arranged pistons; sealing
against the die is effected at a high power, whereas shifting is performed
at a lower power. For this purpose each piston rod is provided with a
small plunger piston.

The counterplaten is connected to the cylinder-platen by three
columns, using split nuts and lock nuts. The return motion of the plunger
crosshead is performed by differential pistons sealed by stuffing boxes,
which may easily be retightened.

Several tool positions in indirect extrusion are illustrated in
Fig. 86.

The advanced container with hollow ram to accomodate die-holder
and die, being attached to the die-carrier, is shown at a. On the plunger
crosshead there is only the tapered ram with screwed-on thrust piece
which – after loading of the billet – advances the container against the
pressure in the two lateral cylinders and tightly closes its bore.

The completed extrusion cycle is shown at b and at c the sawing of
the extrusion after the container has been returned to its end position
and the thrust piece to an intermediate position. During these motions
the shell is torn off the discard and caught in the container. The dovetail
groove in the thrust piece into which the billet engages, serves to entrain
the discard. This groove is slightly tapered so that a light blow with a
hammer suffices to separate discard with die-holder from the thrust
piece. The saw is pivoted about the top column and for direct extrusion
it is dismounted.

Prior to beginning another extrusion cycle the container must be
advanced and retracted once to allow for the ejection of the shell, this
being done with the help of an ejector disk which is placed on the ram
and fits the container bore closely. Thereupon ram and die-carrier are
traveled out, cooled and the die is put in place again.

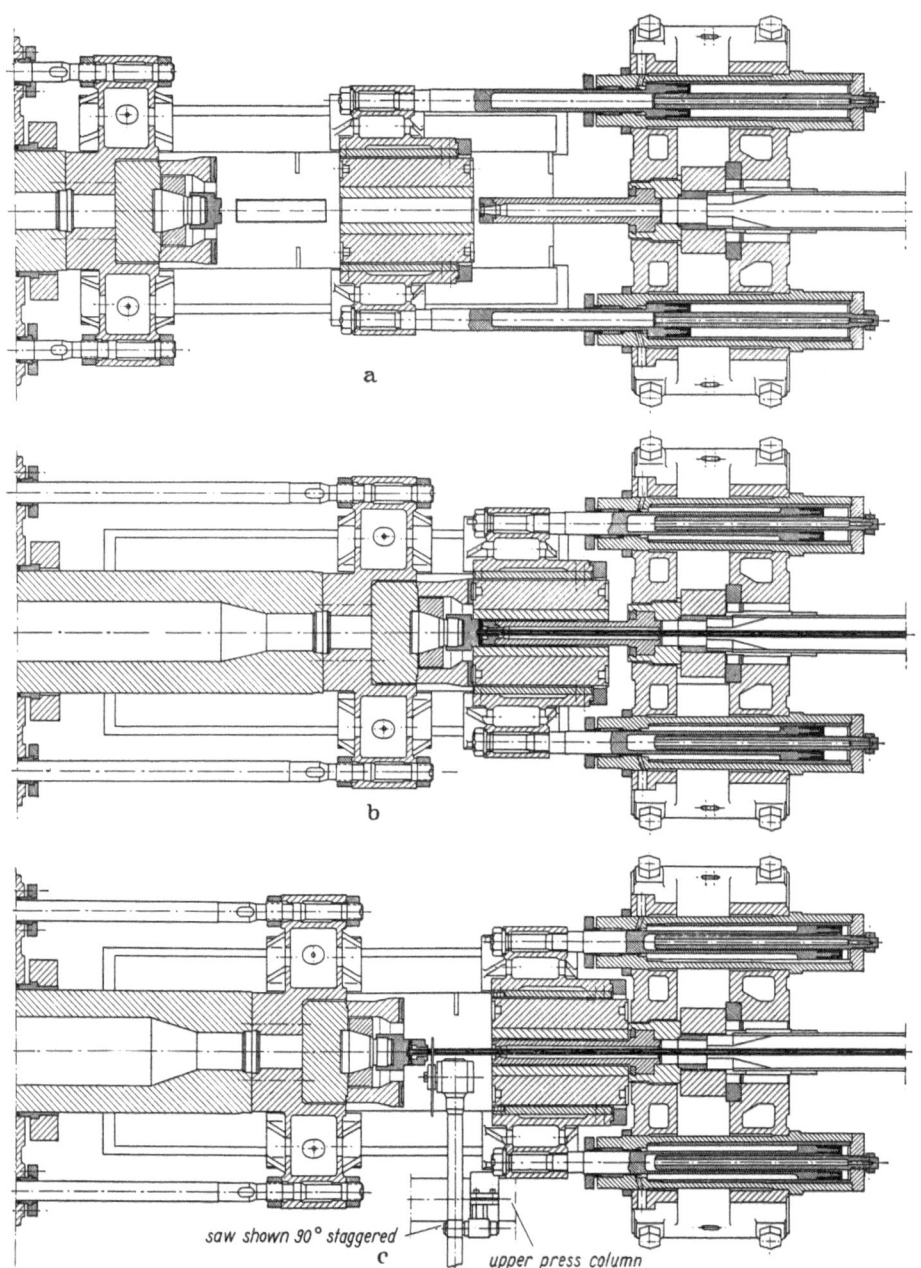

a

b

saw shown 90° staggered

c

upper press column

Fig. 86a–c. Sequence of operations in indirect extrusion in press as per **Fig.** 85

c) Horizontal Combined Rod and Tube Extrusion Presses

Rising demands that had to be met with in extrusion presses when producing tubes, led – as early as about 50 years ago – to the construction of a horizontal combined rod and tube extrusion press, in which the billet is pierced independently of the ram motion. This design also permits of holding the mandrel in the die and of withdrawing it into the ram while loading the press, so as to reduce the stroke of the press. It is true that the device shown in Fig. 74 allows for the mandrel to be shifted too, but in this case the mandrel passes through the die during the extrusion cycle, since its motion depends on that of the ram. The necessity to hold the mandrel stationary during the motion of the ram is, for example, given when tubes or hollow sections of very small inside diameter are to be produced, in which case only the front end of the mandrel tapers to the shape of the hollow section, this being done so as to increase the strength of the mandrel.

One of the earliest combined rod and tube extrusion presses, built in 1909, is illustrated in Fig. 87. Its design is similar to that of the early horizontal four-column rod extrusion presses and was provided with a central through piercing rod. The piercer cylinder was arranged behind the main cylinder and secured to the four extended columns. The main plunger was provided with a sleeve-like extension to allow for the piercer rod to pass through. This extension was sealed at the bottom of the cylinder by a stuffing box and its end served as stop for the two stroke limiting nuts on the piercer rod. At its rear end the piercer rod had a crosshead on which two pullback cylinders, arranged in the horizontal plane on either side of the piercer cylinder, acted. The pullback device for the main plunger was arranged on either side of the main cylinder and transmitted the pullback power on the main plunger crosshead, which was provided with a piercer rod guide, and on which the ram was supported.

A later design of tube extrusion presses is shown in Fig. 88. This press is in three-column construction. Two piercer cylinders b are arranged on either side of the main cylinder a. The piercer plungers c press on the piercer crosshead d, moving in a slot of crosshead e and being guided in it on adjustable gibs. The piercer rod f, into which the mandrel-holder g is screwed, is fixed in the piercer crosshead. The piercer rod moves in the cylindrical bore of the ram-holder, whereas the mandrel-holder is guided in the bore of the ram. The mandrel is screwed into the mandrel-holder and carries at its front end the dummy-block. The wall thickness of the tube to be extruded is easily varied by adjusting the ram with the help of several wedges h. This adjustment affects also the piercer rod which must therefore have a little play in the piercer crosshead. Short differences in the mandrel lengths due to remachining are compensated by spacers i in the piercer crosshead. Guide shoes j sliding on the three columns and being adjusted by shims, serve to center the crosshead and to support its dead weight. The piercer and main plungers have a common pullback device. In order to reduce the overall length of the press, the pullback cylinder is suspended in the main cylinder, for which purpose the main plunger must be bored out correspondingly. The rods k which are attached to the pullback crosshead, are connected with the piercer plungers and provided with split stroke limiting nuts e, which hold the mandrel in a given position in the die when coming into contact with the sleeves m.

A cycle is performed in that – after the billet having been loaded – ram and mandrel are advanced against the billet, when the main cylinder is being filled with low-pressure water from an air-vessel or a high-level reservoir. While this motion is continued, the billet is pierced in the container. The piercing stroke is finished when the mandrel has passed into the die for about 50 to 100 mm and the stroke limiting nuts have made contact with the sleeves. Now the extrusion stroke is performed when the crosshead passes over the stationary piercer crosshead and the ram forces the billet through the annular space formed between mandrel and die. The billet having been extruded, the die-carrier is unlocked to allow for dummy-block and discard to be ejected from the container. During the subsequent return stroke the crosshead is retracted. Having reached its end position, the mandrel is advanced by the piercer plungers and cooled by means of a water- or oil spray. This is followed by another loading operation, after having first retracted the mandrel.

Like in the older types of extrusion presses, the counterplaten is equipped with a wedge-type locking device for the die-carrier which is shifted by hand. The shear arranged behind the platen, cuts in the vertical direction. The bottom blade is pivoted in and out by turning shaft n so as to allow for the shifting of the die-carrier.

A considerably improved design of this type of press is shown in Fig. 89. It is equipped with separate return devices for crosshead and piercer crosshead and an advancing device for the main plunger. This renders possible to advance the ram with the mandrel retracted, whereas up to that time the piercer crosshead with the mandrel was advanced first thus entraining the crosshead. This improvement allows for the billet to be upset in the container prior to piercing and for more accurate wall thicknesses in the tubing, because the mandrel can no longer deflect in the upward direction (see p. 146).

Also in this design the pullback cylinders a and b for the main crosshead and the piercer crosshead are suspended in the main- and piercer cylinders to keep the length of the press to a minimum. The pullback

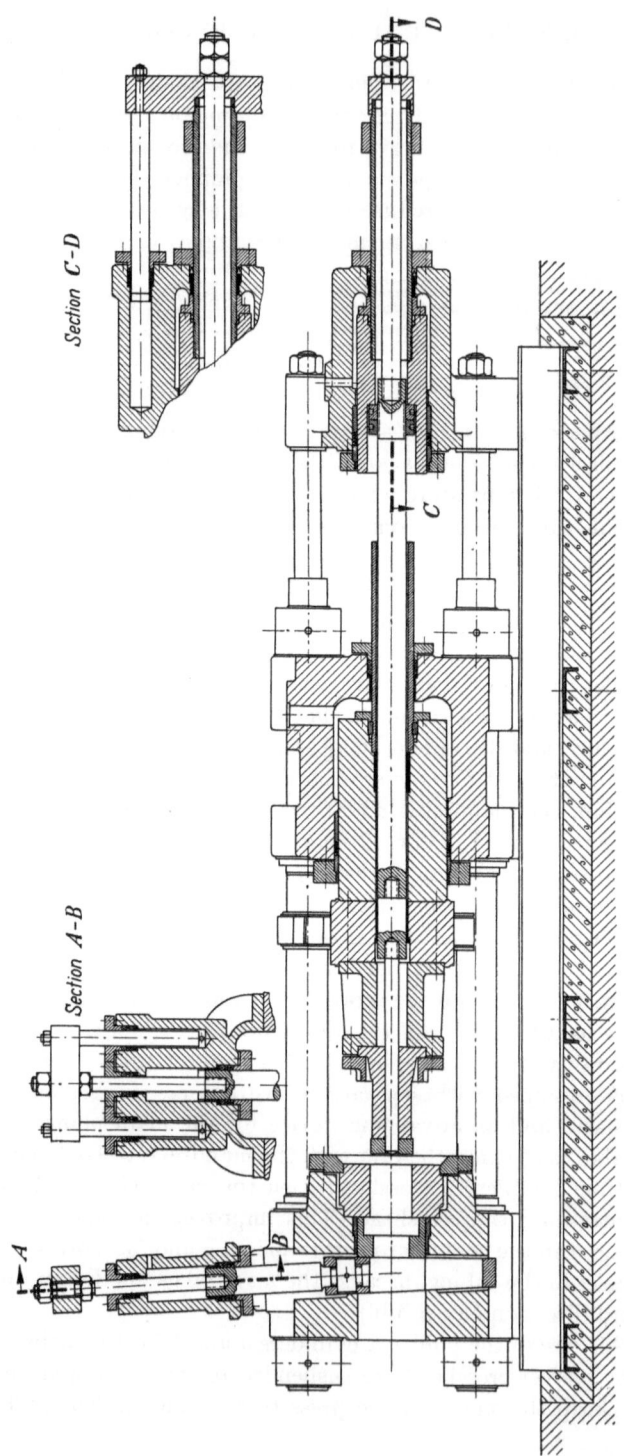

Section C–D

Section A–B

Fig. 87. First combined tube and rod extrusion press of 2,000-ton capacity, built in 1909, with centrally arranged drive of the piercer for independent mandrel motion
(By Hydraulik, Duisburg)

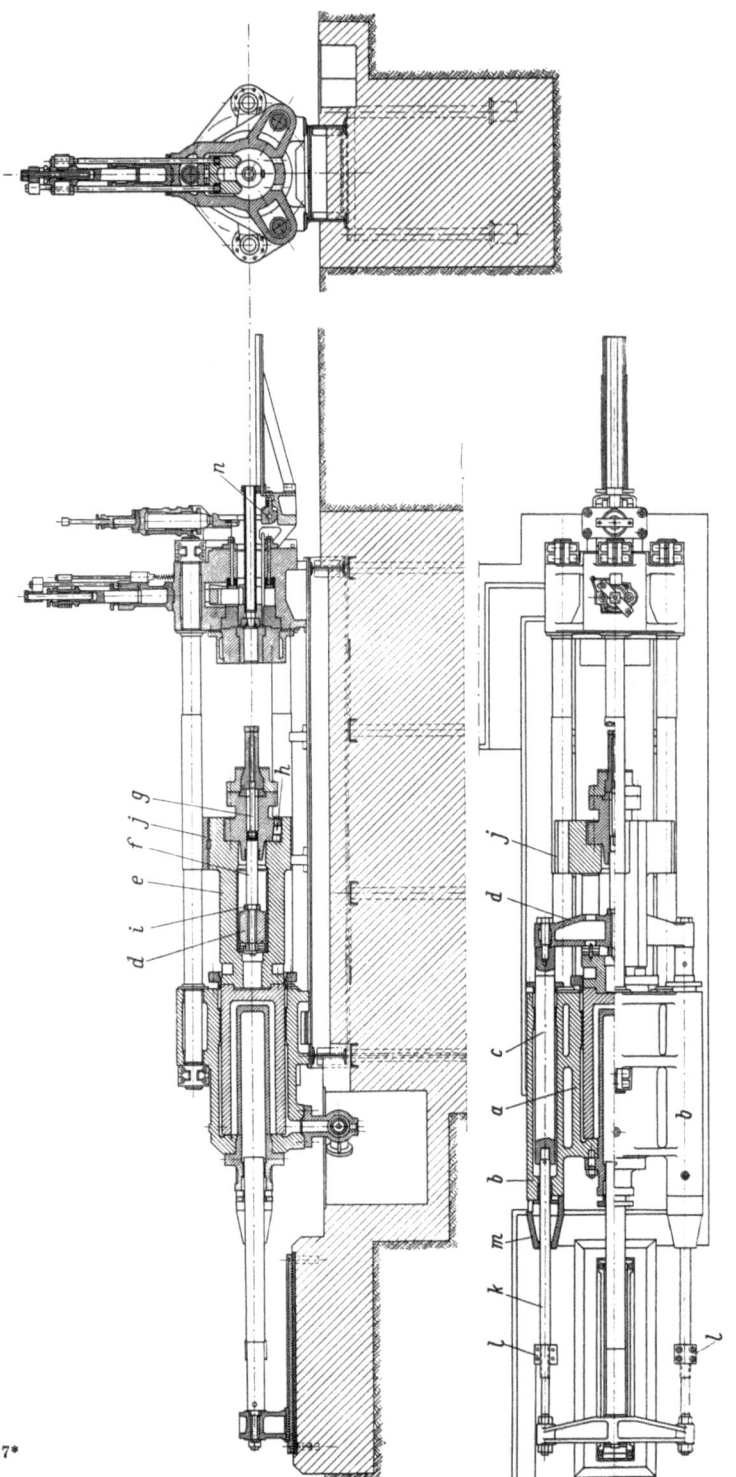

Fig. 88. Combined tube and rod extrusion press in three-column design with laterally arranged drive of the piercer and common pullback device for main ram and piercer mandrel

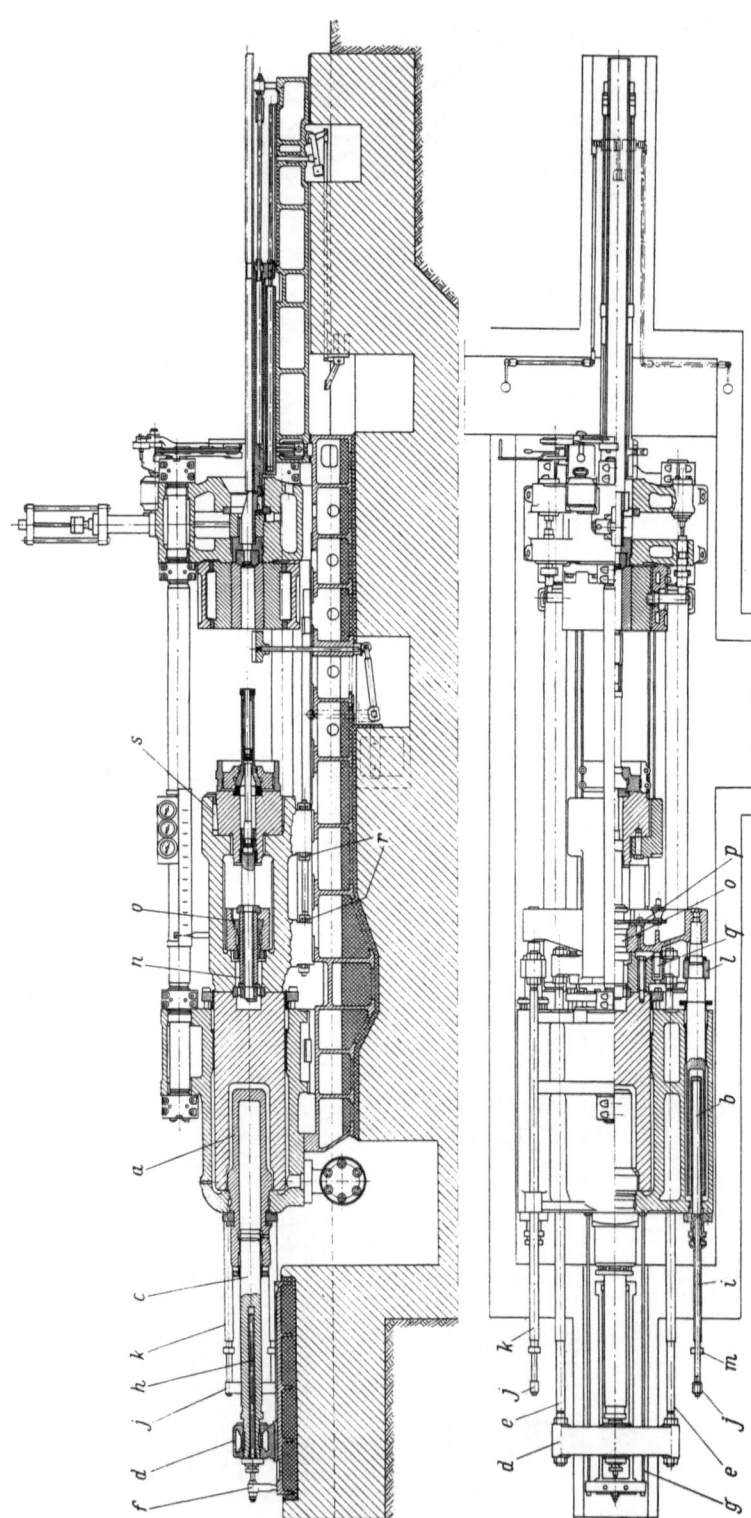

Fig. 80. Combined tube and rod extrusion press in three-column design with laterally arranged drive of the piercer and separate pullback devices for main ram and piercer mandrel
(By Hydraulik, Duisburg)

plunger c presses on the crosshead d which is connected to the main crosshead by two rods e and reliably guided in its travel on a base plate. The end of the base plate carries an abutment f with two tie-rods g to the main cylinder; it serves to support the advance plunger h which moves in a long bore of the pullback plunger c and is sealed by a radially adjustable stuffing box. The pullback plungers i for the piercer device act via the crossheads j on the rods k which are connected to the pullback crossheads l on the piercer plungers. The piercing stroke and/or the position of the mandrel in the die is fixed by the stops m.

Short differences in the mandrel lengths are compensated in the press shown in Fig. 88, by appropriate spacers. If, however, high-tensile billets of short length are extruded which call for a considerable reduction in the length of the mandrel, it is simpler to provide the piercer rod with an adjusting device. It comprises a spindle n and a nut o being rotated in the piercer crosshead by spur-wheel drive p. The pinion is rotated with the help of a detachable crank.

A very practical feature of this press are the stripping cylinders q mounted in the main crosshead, the pistons of which are under constant pressure being somewhat higher than the upsetting power on the piercer mandrel. This ensures that, when upsetting a billet, a clearance of about 100 mm is maintained between the main crosshead and the rear edge of the piercer crosshead, so that upon completion of the extrusion and the die-carrier having been traveled out, discard and dummy-block can be ejected from the container over the stationary mandrel, when the pressure on the main plunger overcomes the power of the stripping pistons. If the edge of the piercer crosshead butted against the main crosshead, when upsetting the billet, provided that the dummy-block be flush with the mandrel, it would not be possible to push a close-fitting dummy-block out of the container because of the limitation of the piercing stroke.

In the end position of the ram the main crosshead rests against the container with the stroke limiting sleeve which is attached to the ram-holder. The thickness of the billet discard is read on a graduated scale; a stroke limiting sliding bar as with the press shown in Fig. 59 and 73, is therefore not required.

Guiding of the main crosshead on the columns affects the alignment of the ram due to the relatively high load on the columns and their tendency to elongate and to deflect during extrusion. With tube extrusion presses it has been general practice to guide the main crosshead on the base frame and to align it with the help of adjusting wedges r. The adjusting wedges s serve for the fine adjustment of the ram axis.

The counterplaten is equipped with a flat sliding bar and a container sealing device similar to the rod extrusion press illustrated in Fig. 67.

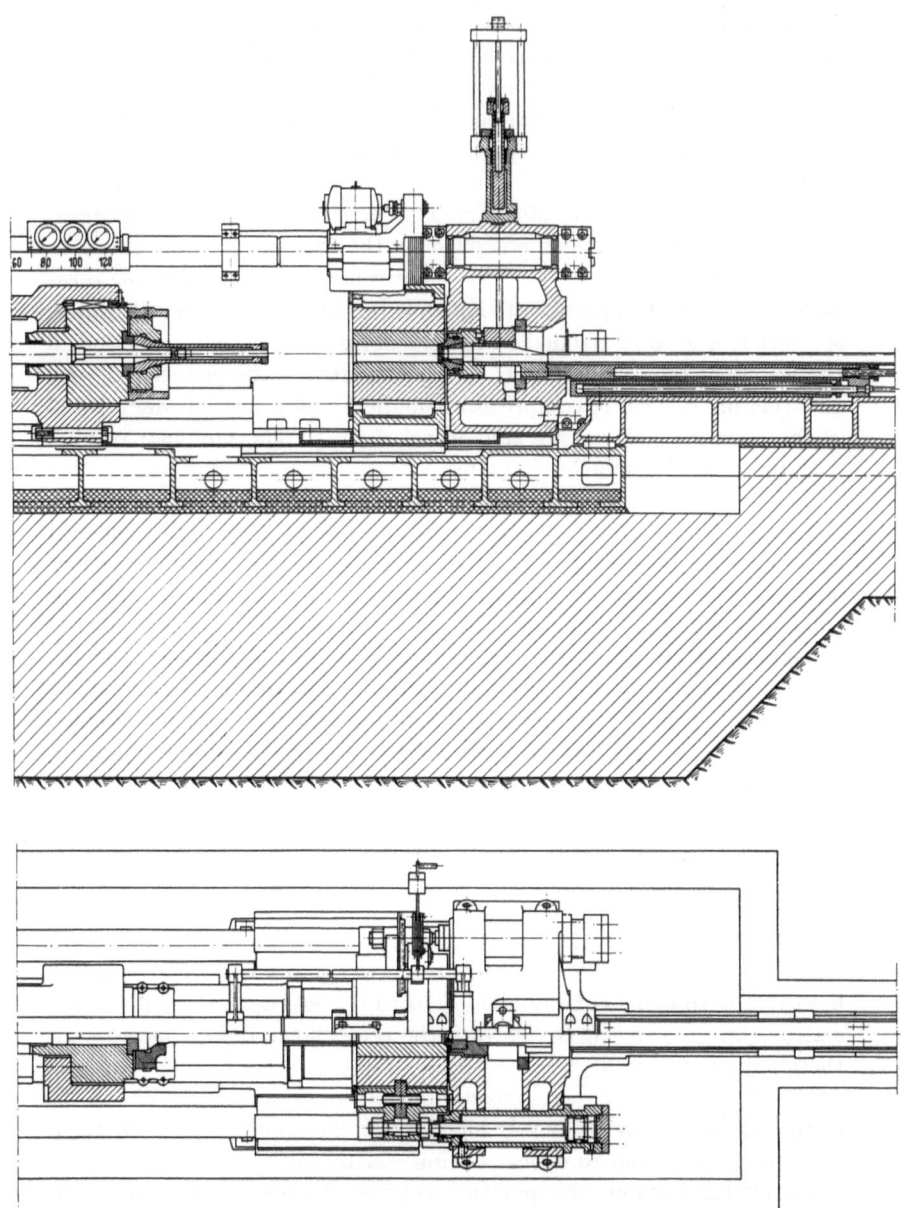

Fig. 90. Combined tube and rod press in three-column design suitable for direct and indirect extrusion
(By Hydraulik, Duisburg)

Tooling arrangement, die-carrier shifting device and saw are as shown in Fig. 66.

The afore-described press may be arranged for indirect extrusion without any great expenditure. For this purpose it must be possible to shift the container for about one ram length which requires long shifting cylinders instead of the sealing device. Fig. 90 illustrates the design of the press. The shifting cylinders are arranged in the horizontal center plane of the counterplaten. The pistons are of the double-acting type and attached to the container-holder which moves on two slideways embracing the columns.

In indirect extrusion it is not possible to cut the extruded rod from the traveled-out die-carrier behind the counterplaten. Separation is performed between the advanced container and the counterplaten with the help of a pendulum saw pivoted about the top column.

A normal sequence of operations when extruding rods by the indirect method, may be followed by reference to Fig. 91. The container in the retracted position is shown at a; behind it there is the die-carrier with the hollow ram carrying the die. The billet is inserted from the front into the container which is then closed by the thrust piece on the main crosshead and forced against the ram. b shows the end position after completion of the extrusion and c shows the sawing of the rod after the container has been retracted fully into its end position and the main crosshead has been traveled back over a given distance. The billet discard with the die is knocked out of the conical or dovetail groove using a hammer. For direct extrusion an additional shear is provided in many cases behind the counterplaten.

The operation of a tube extrusion press as per Fig. 89, may be followed best by reference to a hydraulic circuit diagram, Fig. 92. The idle stroke is not performed by an advance piston, but by low-pressure or filling water of relatively high pressure.

The press has two main controls for pressing and piercing, as well as four secondary controls for container sealing and stripping, die-holder locking, die-carrier advancing and retracting, and for clamping of the billet discard when sawing off the rod. The inlet valves are marked by odd numbers and the outlet valves by even numbers. Each valve is provided with a balance pin (see p. 223) which serves to balance the control shaft when the valve is opened and pressure is admitted to the valve stem. The throttle slide-valves arranged between valves 3 and 4 in the main controls serve to control the speeds of the ram and of the piercing mandrel. Check valves 5 prevent excessive pressures building up in the appertaining cylinders in the event of the outlet valves failing to operate. The valve lift diagrams show the sequence of opening and closing the valves when changing over from one control position to another.

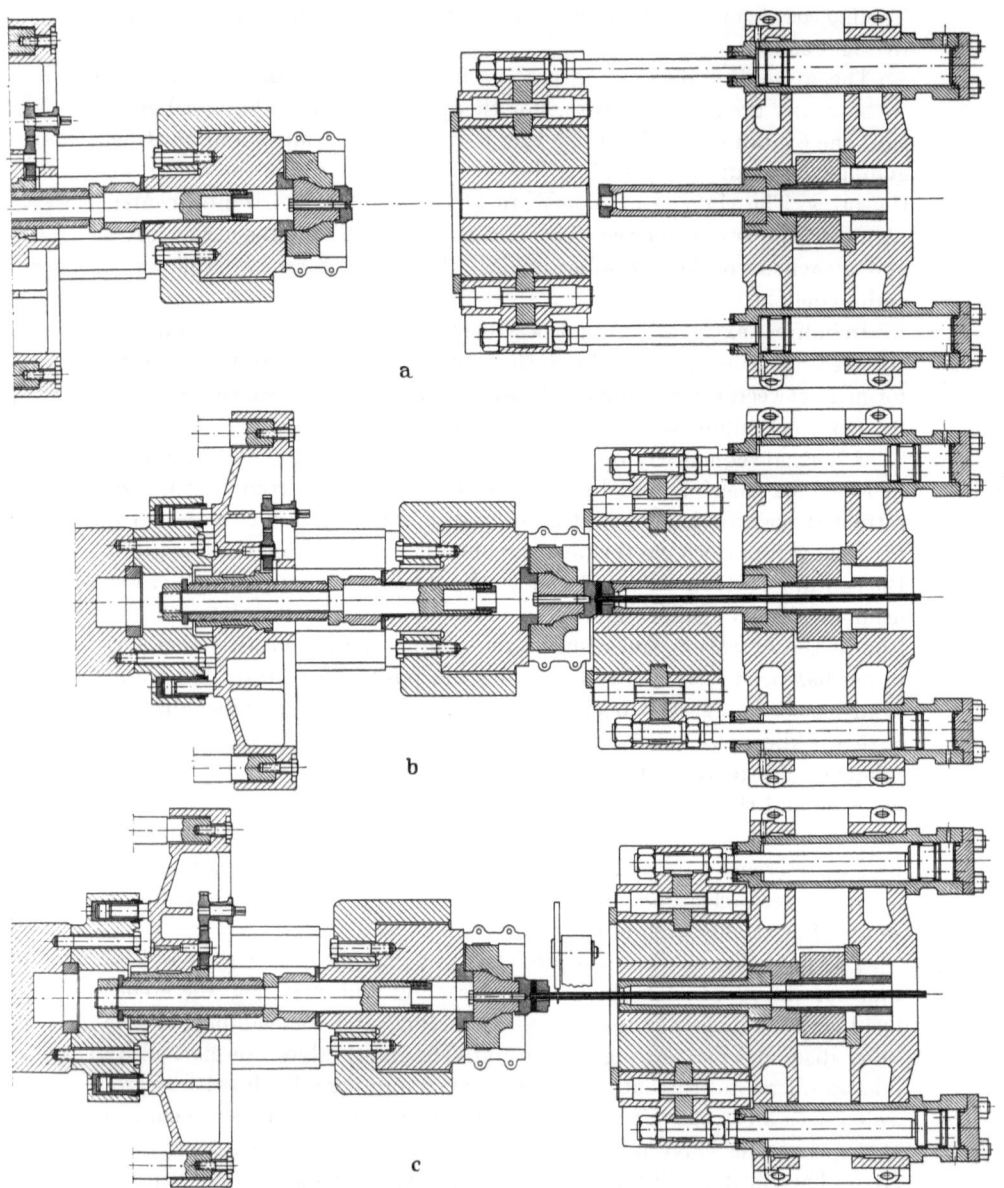

a

b

c

Fig. 91 a–c. Sequence of operations in indirect rod extrusion in the combined tube
and rod press as per Fig. 90

At the beginning of an extrusion cycle, i. e. after the billet has been
loaded, the lever of the pressing control is shifted from *stop* to *advance*.
The low-pressure water, being under a pressure of 6 to 8 atmospheres,

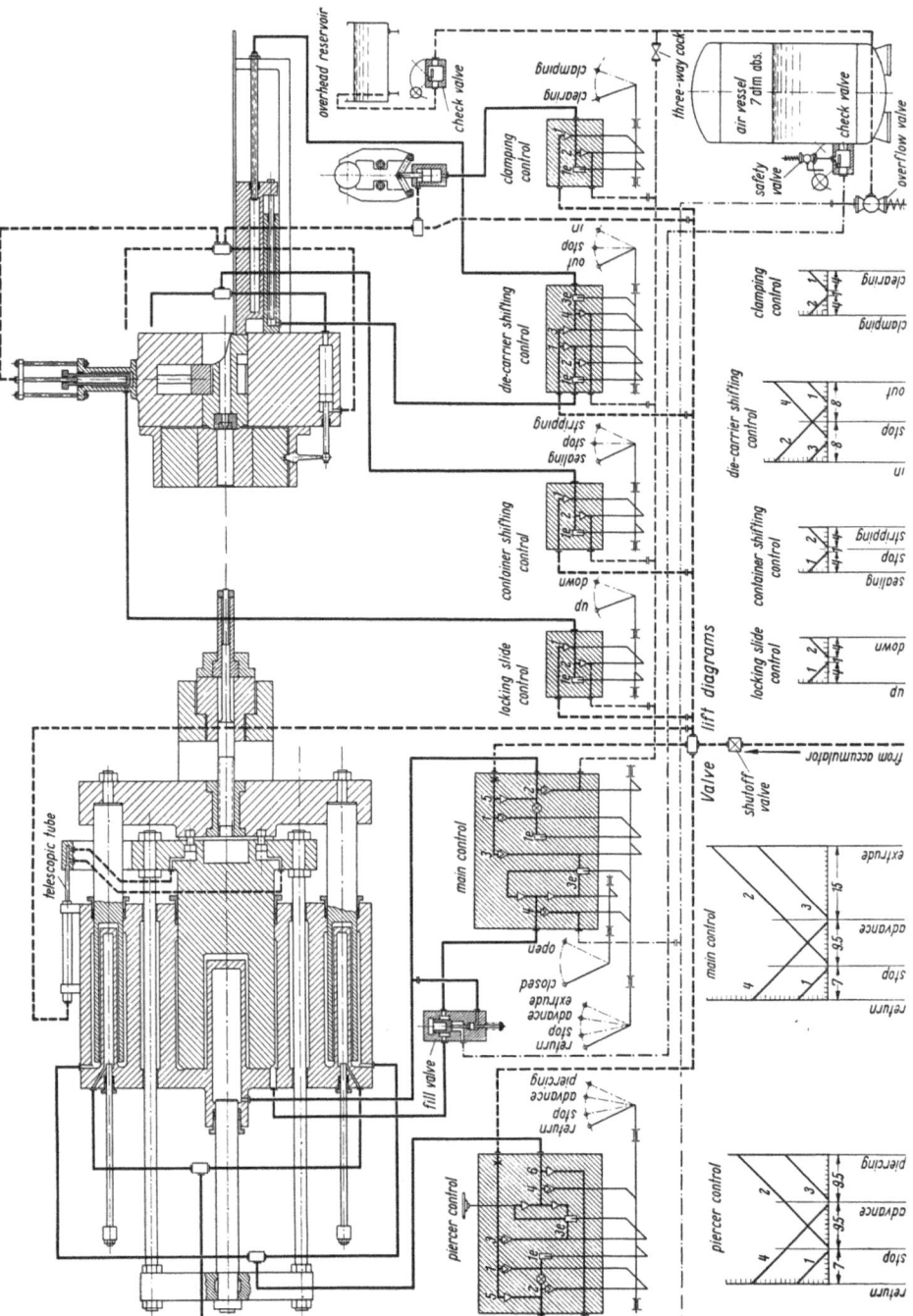

Fig. 92. Hydraulic circuit diagram for a combined tube and rod extrusion press as per Fig. 89

flows from the air-vessel through the filler valve which acts as a check valve, into the main cylinder and advances the central plunger in an idle stroke when the water which is forced out of the pullback cylinder, flows through the opened valve 2 into the high-level reservoir. It is not fed into the air-vessel because of the greater counterpressure prevailing in there. While advancing, the plunger entrains the piercer crosshead. Consequently, the piercer cylinders must be filled so as to prevent a vacuum for which purpose the suction valve 6 is provided in the piercer control. A small clearance is maintained between the plunger and the piercer crosshead by the pistons in the piercer crosshead which are under constant accumulator pressure. This pressure is overcome only when – after completion of the working stroke – the dummy-block which loosely fits the mandrel, has to be pushed out of the container.

The dummy-block, which is flush with the mandrel, having come to contact the billet, the latter is upset. For this operation a small quantity of pressure water is fed into the main cylinder by momentarily shifting the lever of the press control to *pressing*, thus opening the inlet valve; thereupon the lever is moved back to *advance*. For the piercing of the billet the lever of the piercer control is shifted from *stop* to *piercing*. The pressure water flows from the accumulator through the opened inlet valve 3 into the two piercer cylinders, while the water which is forced out of the pullback cylinder is fed through the opened return-water valve 2 into the air-vessel. The *advance* position in the valve lift diagram has been crossed. It is only used when the mandrel is traveled out of the ram for cooling or changing. Prior to beginning the piercing operation the lever of the press control has to be shortly shifted to *return* and subsequently to *stop*, thereby withdrawing the ram slightly to allow for the billet to elongate during piercing. The end position of the piercing mandrel in the die is fixed by stops on the pullback rods of the piercer. Extrusion of the billet is initiated by opening valves 2 and 3 of the press control and shifting the hand lever from *stop* to *pressing*.

The working stroke having been completed, the ram is withdrawn slightly and the container stripped from the die so that the locking bar of the die-holder may be unlocked and lifted. Then the dummy-block is pushed out of the container by the ram and the die-carrier is retracted. In the end position of the die-carrier the discard is clamped, whereupon the stroke limitation for the die-carrier is disengaged by actuating a foot pedal to allow for a limited excess stroke to be performed to permit of sawing off the extruded rod. When pushing out the dummy-block, the ram-holder moves against the container with its stroke limiting collar. Upon ejection of the shell, container, die, mandrel and fresh dummy-block are lubricated and the plungers are moved back into the starting position. All the controlled flows of water can be followed easily by reference to the valve lift diagrams.

When the plunger moves back, the water contained in the main cylinder must be forced back through the filler valve into the air-vessel. The filler valve is opened by a stem and a piston, the cylinder of which is connected to the return pipe (see p. 78). As the force of the stem is not sufficient to open the filler valve against the accumulator pressure, a small relief valve *4* is arranged in the press control, by means of which he pressure in the main cylinder is relieved prior to beginning the return motion. The quantity of water which flows back into the air-vessel is always larger than that previously withdrawn, the excess comprising of that quantity of water that has been supplied by the accumulator to perform the working strokes. The air-vessel is therefore provided with an overflow valve opening automatically at maximum pressure and allowing the excess water to flow back into the high-level reservoir.

Fig. 93 shows an improved tube extrusion press with central piercer bar as per Fig. 87. The counterplaten which accomodates a die-carrier shifting device and a wedge for locking the die-carrier as already des-

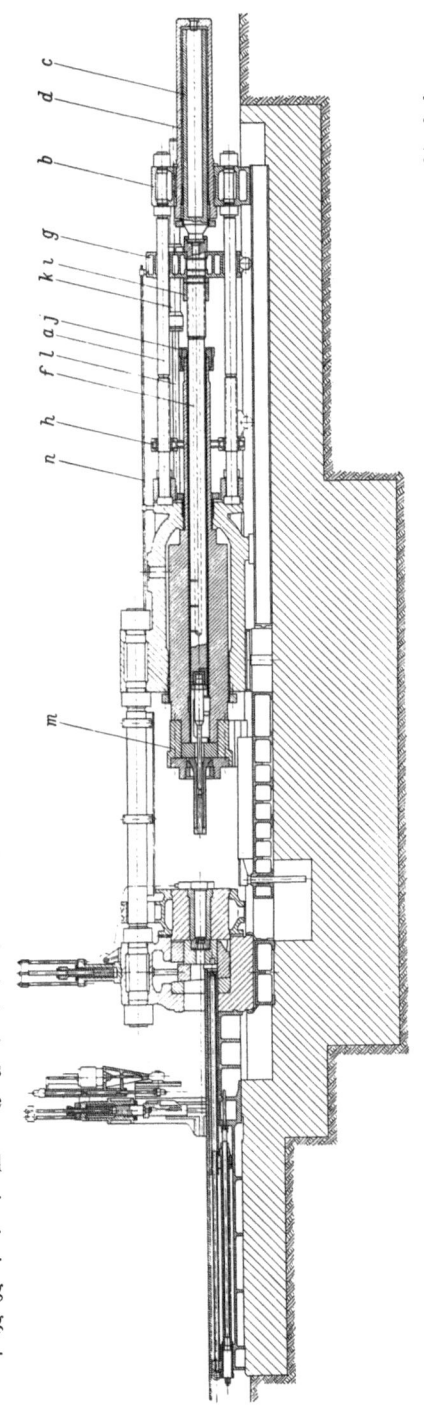

Fig. 93. Combined tube and rod extrusion press in three-column design with centrally driven piercer, die-carrier shifting device and combined shear and saw on the shifting bed (By Schloemann, Düsseldorf)

cribed in this book, is connected to the cylinder-platen by three columns. The bottom of the cylinder carries two columns a which are positively connected to crosshead b being mounted on the extension of the base frame, and accomodating the piercer cylinder d. The piercer plunger c, being bored out to keep its own weight to a minimum, presses against the piercer bar f. The latter is supported by a crosshead g which moves on the base frame and whose stroke may be varied as required by means of the two stroke limiting nuts h on the threads of the columns a. The stroke of the mandrel relative to the ram may also be varied. For this purpose the nut i which comes into contact with the plunger extension, is adjusted on the thread of the piercer bar. The rear limitation of the stroke of the piercer bar is ensured by two lateral drag links which connect the crosshead g to a crosshead j on the plunger extension and which also serve to compensate differences in the lengths of the mandrels. The piercer device may be disengaged by removing the drag links.

The piercer bar is retracted by two cylinders k which are arranged in the crosshead g and move over the stationary plungers l. The pullback device for the main plunger is not shown in Fig. 93. The plungers employed for this purpose can either press against crosshead j from the cylinder-platen side or, for example, pull at the main crosshead m through a reversing linkage.

The piercer bar is guided within the main plunger in a long bronze bush. The sliding surfaces are lubricated with grease which is supplied through the inner bore of the piercer bar. A rod n being attached to the crosshead g, indicates the travel of the piercing mandrel.

Fig. 94 illustrates the construction of the cylinder end of a tube extrusion press with through piercing bar and an improved mandrel positioning device. Mandrel, mandrel-holder and mandrel-rod are guided in the ram and the ram-holder; they may be adjusted for centering in the die both radially and axially. The radial adjustment of the mandrel is effected by aligning the ram-holder by means of setscrews. In order to render possible axial adjustment of the mandrel to suit different billet lengths as with the press shown in Fig. 93, the mandrel-rod is connected through a claw coupling to an adjusting sleeve being rotated in the piercer bar head which is designed in the form of a threaded bush. Rotating is effected through a long shaft lying in the piercer bar, by a worm drive arranged in the crosshead of the piercer plunger; the worm drive is operated by hand or by a small pushbutton controlled motor. The shaft rotates the nut via a transmission gear.

With tube extrusion presses, in which only hollow billets are used to ensure very smooth inner surfaces of the tubes – this being frequently done in light metal works – a relatively low power is required to advance the mandrel. Therefore these presses are not provided with a piercer but with a mandrel shifting device similar to the former; radial adjustment of the mandrel is not required because the hollow billets lie concentrically and therefore need not be upset. A tube press with mandrel shifting device is shown in Fig. 95. In this case the piercer bar may be used as the shifting cylinder. Consequently, the shifting plunger is stationary and set in a crosshead being mounted at the end of the base frame extension. This crosshead is connected to the main cylinder by two columns which are not shown in the drawing. The stroke of the rear piercer bar crosshead, which is guided on the base frame, is limited by collars on these columns. A nut in the piercer bar crosshead, rotated by a worm drive,

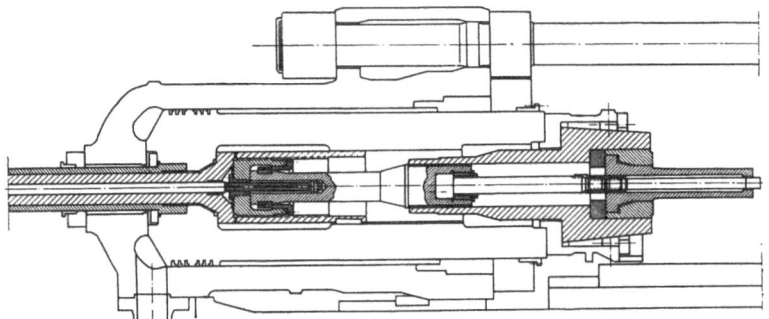

Fig. 94. Mandrel positioning device to adjust variable mandrel lengths

serves to adjust the piercer bar so as to be able to set the position of the mandrel in the die according to the various mandrel lengths. At the side of the columns there are arranged two pullback cylinders which are supported on the cylinder-platen and whose plungers push against the piercer bar crosshead. Similarly two pullback plungers act on the crosshead on the extension for the main plunger.

Fig. 96 shows an older light metal tube extrusion press, built in 1936/37, in frame construction, being the first press to be equipped with a horizontal die slide and a moving container, even before using it in rod extrusion presses (see p. 83).

The press frame a is carried on transversely set base plates b having its fixed abutment at the cylinder end. At the container end the frame is loosely supported in a manner which permits expanding due to pressure and temperature differences. Plunger d, being sealed by stuffing boxes, slides in the main cylinder c. Crosshead e with two pistons f, being arranged in an inclined center plane, and appropriate cylinders are provided for the return motion. The pullback crosshead is carried on adjustable guide-bars on bed g which is separate from the frame. This bed also serves to guide the container-holder and is not affected by the expansions of the frame.

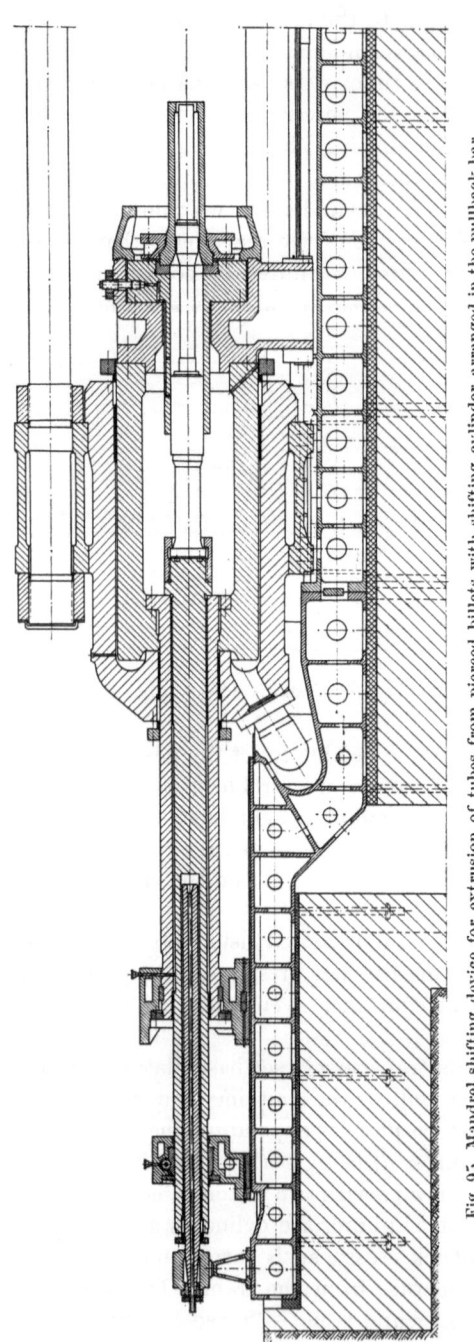

Fig. 95. Mandrel shifting device for extrusion of tubes from pierced billets with shifting cylinder arranged in the pullback bar

Plunger d transmits the extrusion power via a thrust plate to the ram, which is centered by setscrews and in which the mandrel-holder with attached mandrel is guided. The mandrel-bar h moves in the plunger extension i which passes through the cylinder bottom; it is attached to a shifting piston j which, however, remains inoperative as long as it is retained in its cylinder by nut k. This cylinder l is set in the driving crosshead m for the mandrel-bar. The crosshead is moved back and forth by two plungers n and o each. The pullback cylinders are supported on the frame, whereas the piercer cylinders are inserted in crosshead p which is connected to the frame by two tie-rods q. If the driving crosshead m comes to rest against the stroke limiting nuts r, the mandrel is held stationary in the die. If the mandrel is to float during extrusion, nut k has to be released on the shifting piston j so as to enable the mandrel to perform a relative stroke. The operative power of pistons n is relatively small as hollow billets are extruded in this press.

The horizontal slide is moved by a two-armed lever

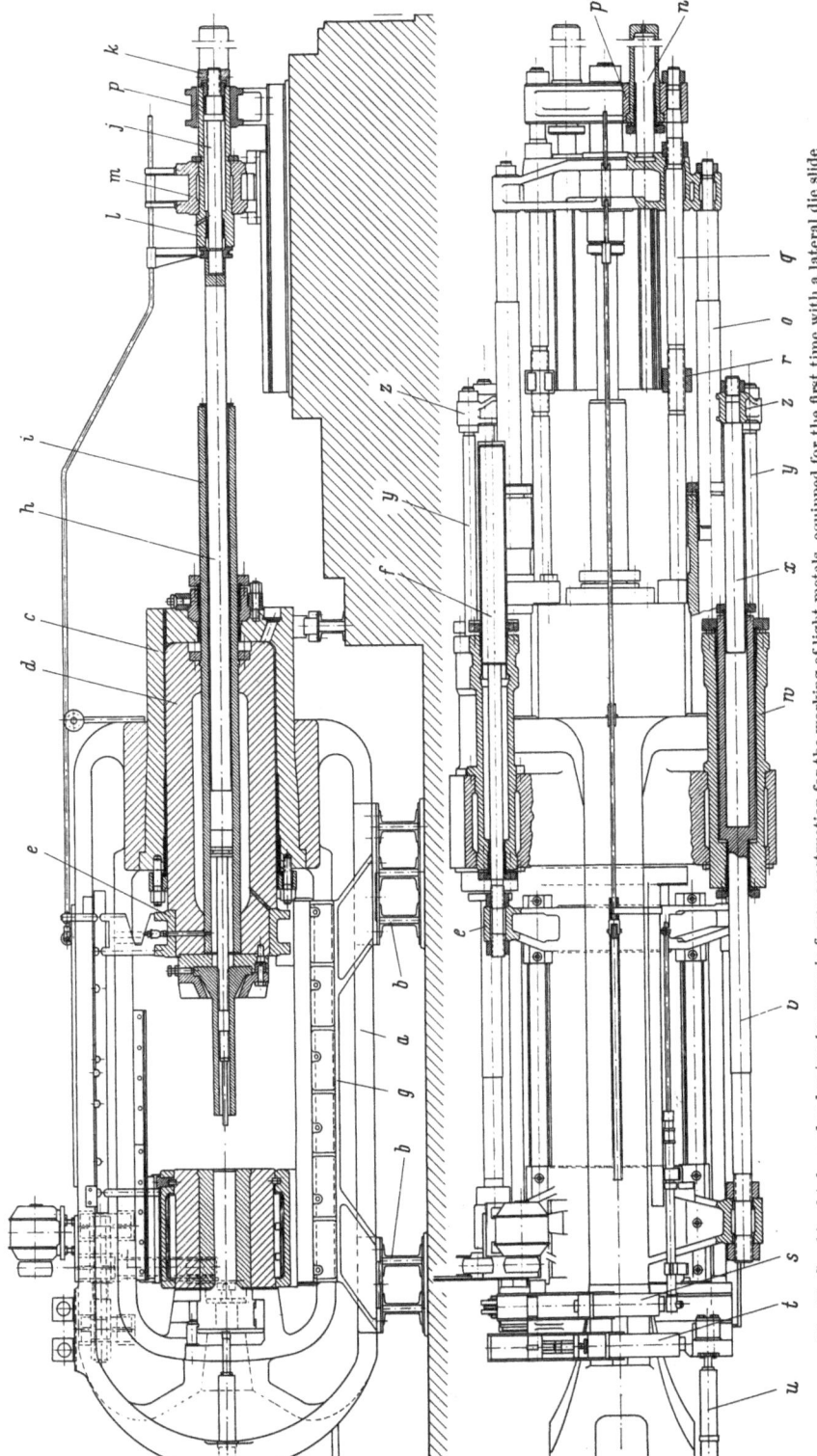

Fig. 96. Combined tube and rod extrusion press in frame construction for the working of light metals, equipped for the first time with a lateral die slide
(By Krupp Grusonwerk, Magdeburg)

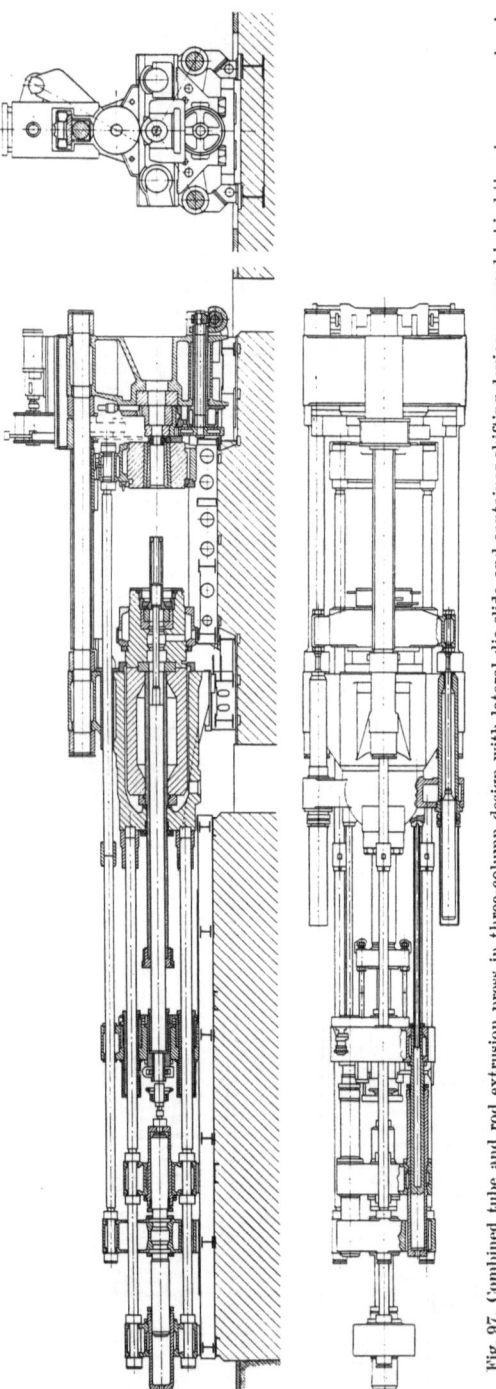

Fig. 97. Combined tube and rod extrusion press in three-column design with lateral die slide and container shifting device arranged behind the piercer crosshead (By Schloemann, Düsseldorf)

which is operated by a piston. The appertaining cylinder s is fixed on the frame. At its side there is another cylinder t, the piston of which rotates a shaft by which a saw is swung in and out. This shaft serves also as pivot for the two-armed lever. The saw blade is operated by an electric motor and V-belt pulleys.

The guide for the horizontal slide is fitted with adjustable gibs. The die having been moved out of the press, it is easily pushed out by a small double-acting piston of cylinder u.

The container-holder is provided with gas-burners to heat the container to working temperature. It is shifted by rods v, the enlarged ends of which move in cylinders w and which withdraw the container from the die, i. e. bring the former into the loading position. To effect the advance motion, the enlarged ends of rods v are designed in the form of cylinders which slide over the plungers x. The latter are supported on a short crosshead z each, being connected to the frame by rods y.

An advanced tube and rod extrusion press in three-column construction with horizontal die slide is illustrated in Fig. 97 and 98. The die slide is brought into its various working positions by a drive unit arranged behind the counterplaten. Extruding having been completed, the die is cleared of the slide during the return motion of the container, so that

Fig. 98. Shop photo of press as per Fig. 97, view of the delivery end

the product may be cut off close behind the die with the help of a sliding saw.

The container shifting cylinders are arranged in two crossheads supported on the main cylinder bottom by rods. The front one of these two crossheads also accomodates the piercer cylinders, the plungers of which act on the piercer crosshead and being designed to form the piercer pullback cylinders. The plungers of the piercer pullback cylinders rest on the bottom of the main cylinder. The piercer crosshead accomodates a handwheel operated device for axial and radial adjustment of the mandrel.

Fig. 99 shows an oil-hydraulic tube press of the frame type in which light metal billets are extruded. On the right-hand side of the frame is arranged the main cylinder with a double-acting piston and a piston extension passing through the bottom of the cylinder. The piercer bar moves in the cylinder axis. The piercer cylinder is also provided with a double-acting piston; it is fixed to the main cylinder by two columns arranged in the horizontal center plane. The piercer stroke is limited by stops on the columns. Piercer bar and piercer piston are connected to

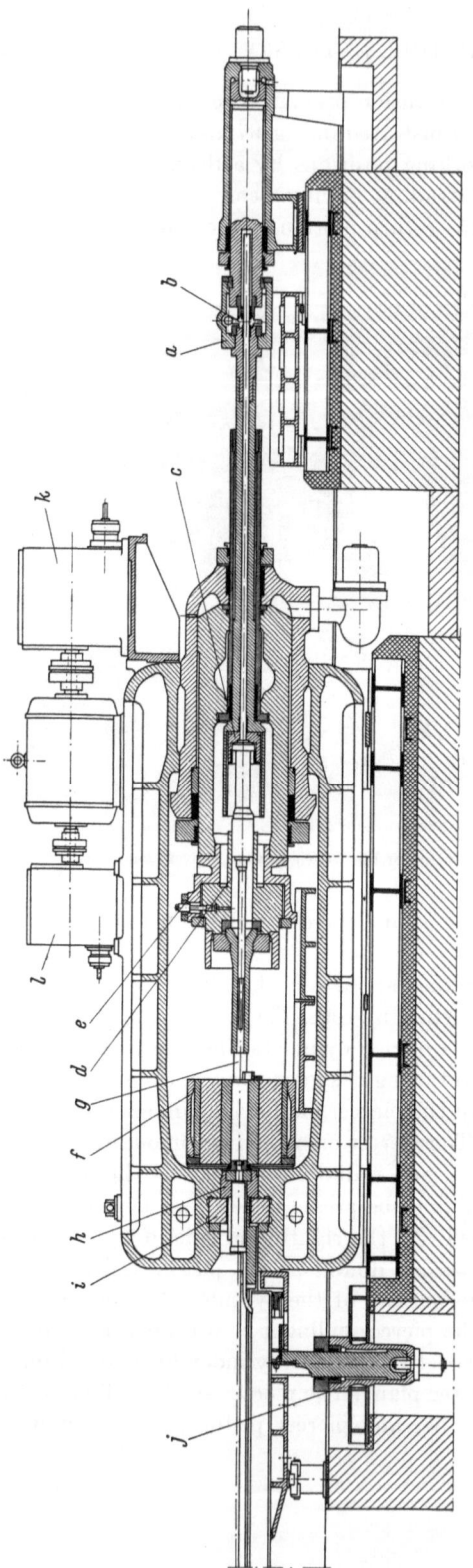

Fig. 99. Oil-hydraulic tube and rod extrusion press in frame construction with central piercer and built-on pumps (By Hydraulik, Duisburg)

each other by a slide a, being guided in its travel on a bed and accommodating the worm drive b for a mandrel adjusting device. In its end position the adjusting bush c comes to rest against the tube-shaped extension d of the ram-holder (see Fig. 94). The two pistons are thus coupled and the extrusion power is increased by that of the piercer. The main ram crosshead e is guided on a bed which is separated from the base frame of the press. The container-holder f whose shifting cylinders and pistons are arranged at the side of the main cylinder, slides on this bed, too. The pistons are of the double-acting type and the piston rods g extended up to the container-holder. The billet is loaded between container and die to keep the stroke of the main piston to a minimum.

Die and die-holder are set in a die-carrier h which is locked by a flat sliding bar i and shifted by a piston. The hydraulically operated shear for cutting the billet discard is arranged in the shifting bed. The shear operating cylinder j is installed under floor level and rests on a structural frame; its positive connection to the die-carrier is established by a swivel arm prior to cutting.

The two variable-delivery axial piston pumps k and l for the motions of the ram and the piercing mandrel are driven by a common motor and arranged on the top of the frame directly above the main cylinder. The oil collector tank and two geared pumps for the auxiliary drives are arranged in a pit below floor level. The auxiliary drive units operate the container and die-carrier shifting devices, the shear, the locking mechanism of the die-carrier and the control gears of the two axial piston pumps.

The design of oil-hydraulic tube extrusion presses is simpler and cheaper than that of water-hydraulic presses due to the fact that double-acting pistons are employed in the former. They have found wide application in light metal works since the ram speeds required in such works are very low and in most cases less than 25 mm/sec., thus requiring small horsepowers only. A further plus is that pumps and motors may be arranged in close proximity of the main cylinder (see pp. 203 and 208).

d) Vertical Combined Rod and Tube Extrusion Presses

Application of the early vertical tube extrusion presses was confined to the manufacture of small tubes of up to about 20 mm diameter in heavy metals. They were designed for a pressure capacity of 300 tons and provided with a gently curved runout table for a tube length of 5 to 6 m. The vertical press was preferred to the horizontal type because of ease of access and higher efficiency.

A drawback of these presess is that the runout table is arranged in a pit (Fig. 100). If the pit is to be obviated, the press has to be arranged on a heavy platform. The advantage of little space requirement is confronted with the disadvantage that either the tubes or the billets have to be lifted after extrusion is completed or prior to beginning the extrusion cycle respectively. Due to the curvature of the runout tray the tubes must be straightened. It is for this reason that vertical presses are hardly ever used for manufacturing sections. In this case the horizontal press is superior because the section is extruded in straight lengths and moreover the length is not limited.

The design of the early vertical tube extrusion presses was simple and cheap. They were of the down-stroking type and built in frame construction. No special piercing equipment was provided since the billets extruded were only small ones. The mandrel was rigidly inserted in the extrusion ram and extrusion is performed on the "without shell" method, when the ram has only a little clearance in the container. The heavy metal

billets are cut from round bars, machined and provided with a central
hole so that very close tolerances in the wall thicknesses are obtained –
this being also due to the exact guiding of the movable crosshead in the
frame. The excellent surface finish of the tubes is due to the use of machined billets. Simpler operation makes up for the expenditure involved in

Fig. 100. Vertical tube presses (one 300-ton and two 600-ton presses). Courtesy of VDM Nürnberg
(By Hydraulik, Duisburg)

machining the billets and permits of performing about 90 to 120 extrusions per hour.

Vertical presses therefore found a very wide application and the next
step was to develop them for higher capacities of 600 to 800 tons, suitable for the extrusion of tubes of about 50 to 60 mm diameter. In many
cases these larger presses were provided with piercing equipment and
also designed to allow the use of dummy-blocks to extrude tubes on the
"with shell" method, thus leading to a wide variety of designs. A few
vertical presses were even built for capacities of 1,000 to 2,500 tons. With
these heavy presses, however, the output is lower than that of horizontal
presses, this being due to the shorter tube lengths produced, so that
the main field of application of vertical presses is confined to a capacity
range of 600 to 800 tons. The tubes extruded on such presses are most

economically drawn to the required smaller dimensions on multiple draw benches.

Fig. 101 shows the simplest construction of a vertical 600-ton tube press. The cast steel frame has a main cylinder which is mounted by inserting it from the top and lowering it so far that the split ring a may be placed around it. The cylinder is secured in the frame by flange b. The hollow chill cast plunger is guided in a bronze bush and sealed by a radially adjustable stuffing box. The plunger base is provided with a throttle bolt which closes the connecting bore in the cylinder leaving open a small slit only, and preventing the plunger from crashing against the bottom at the end of its stroke.

The extrusion power is transmitted from the plunger to the movable crosshead which slides in the frame on a circular or vee-shaped guide. Preference is often given to the circular guide as this can be made with highest precision on a boring mill in exact alignment with the cylinder bore. The guide gibs on the crosshead are adjusted by shims or wedge bolts. Two pullback cylinders are provided in the bottom part of the frame which ensure the return motion of the crosshead. The pullback plungers press on brackets being bolted to the crosshead.

Container and container-holder are rigidly attached to the container support d by four bolts c. This support is aligned to the center of the press by four lateral setscrews (not shown in the sketch) and is provided with a U-shaped slot, open to the outside, through which the die slide e which carries the die, is passed. If the die is to be removed, first the four bolts on the container-holder are released, whereupon the die slide is pulled out of the support manually with the help of a rod and the die which has dropped off, is picked up with a hook.

The container is symmetrical and may be used from either end. It is centered to the container support in an annular groove. The container is held in the holder by a bayonet lock. When dismounting the container, the crosshead is lowered until the stroke limiting collar f makes contact. Then the clamping bolts are screwed in tap-holes specially provided for this purpose in the container, and with the crosshead moving up the container is lifted out. The container-holder is provided with a resistance heating system to heat the relatively small container. Its outer jacket is well insulated to prevent thermal losses.

Utilization of this press is confined to the making of tubes which drop on to a curved runout table h being made of a heavy channel iron; therefore their shape is curved, too. The tubes are picked up with a hand-tongs and straightened in the horizontal straight part of the runout table by slightly pressing them. The extrusion mandrel is rigidly inserted in the ram which has only a few tenths of a millimeter clearance in the container bore. Consequently, no shell is formed during extrusion

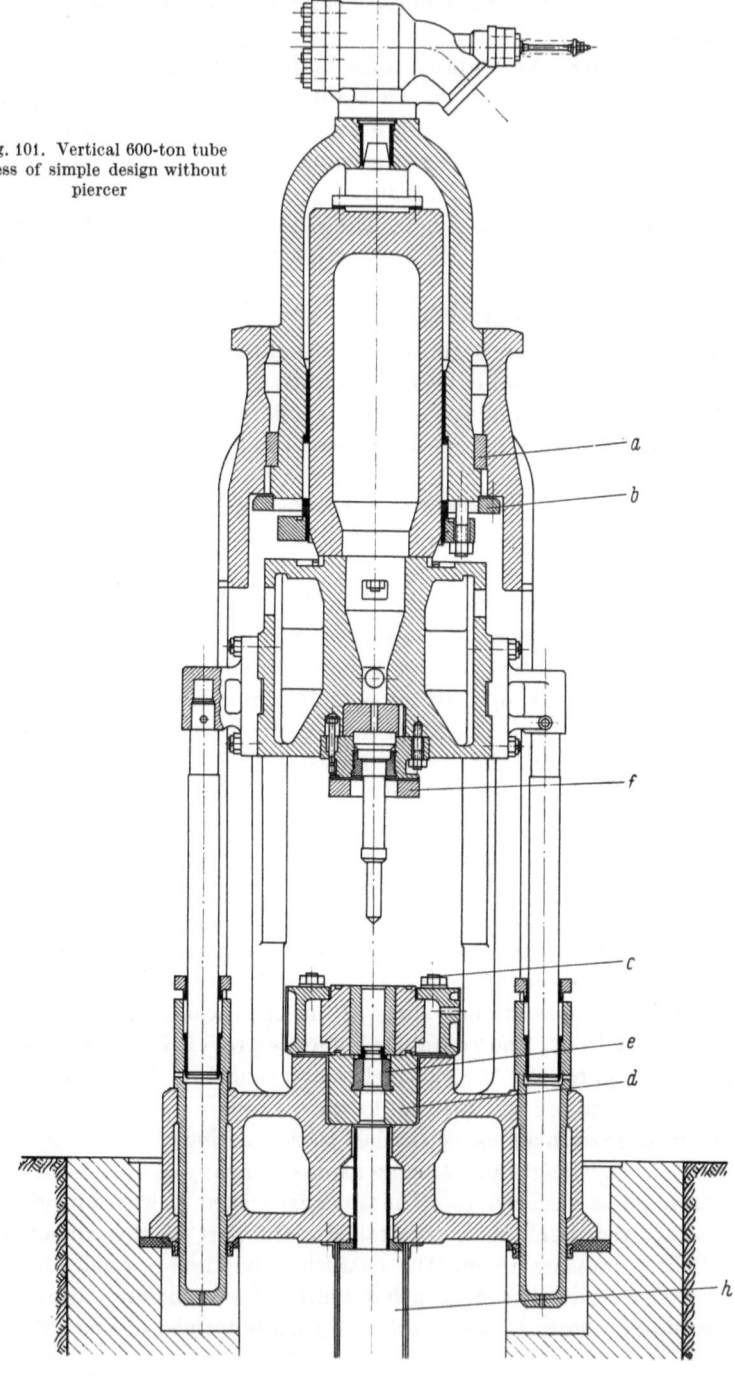

Fig. 101. Vertical 600-ton tube press of simple design without piercer

and the billet discard consists of a thin disk of 2 to 3 mm thickness only.

The tube is separated from the discard by means of a shearing mandrel. This tool consists of a longitudinally slit sleeve, as shown in Fig. 102, which is swung about an axle at the frame into the center of the press and embraces the mandrel. The swivel arm of the sleeve is axially movable on its axle and follows the ram motion. The shearing mandrel, whose diameter is only slightly smaller than the outside diameter of the tube and which is centered in the inside diameter of the tube by a cone, having pierced through the discard, the tube drops off. When the crosshead is lifted, the discard is caught on the shearing mandrel which for this purpose is provided with a small eccentric relief. It is removed from the shearing mandrel by blows with a hammer after the sleeve has been swung out of the press.

In view of the fact that the use of hollow bored billets increases the cost price of the tubes – this only paying off when dealing with alloys which on account of their difficulty in extrusion give rise to excessive mandrel wear – piercing in many cases is carried out by the extrusion mandrel. The press illustrated in Fig. 101, is, however, not suitable for this, as in this case the pressure water consumption of the main plunger would be needed to overcome the relatively low resistance to piercing. Vertical tube presses, in which the billets are to be pierced, are therefore provided with special piercing equipment. Fig. 103 shows an example of this type of press in which two additional piercer cylinders, lying in the same axis as the pullback cylinders, are arranged in the top part of the frame.

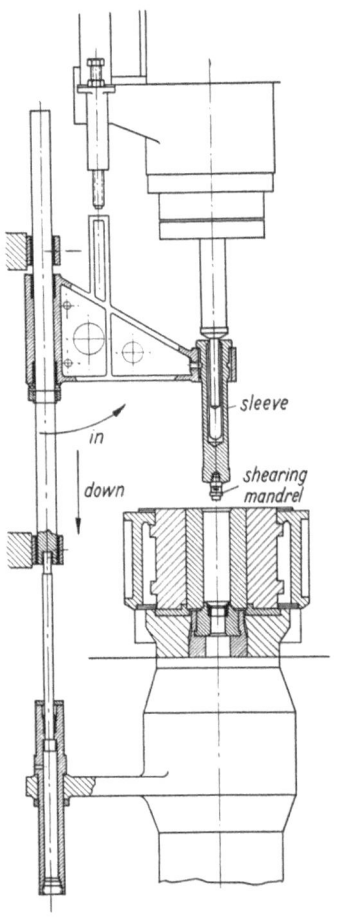

Fig. 102. Device for separation of tube from discard for a vertical tube press

A hydraulic die lifting device is arranged underneath the press frame which is used when occasionally extruding solid sections on the tube press. With the die being lifted, the discard may easily be separated from the extruded section with the help of a chisel, as the cross-sections of

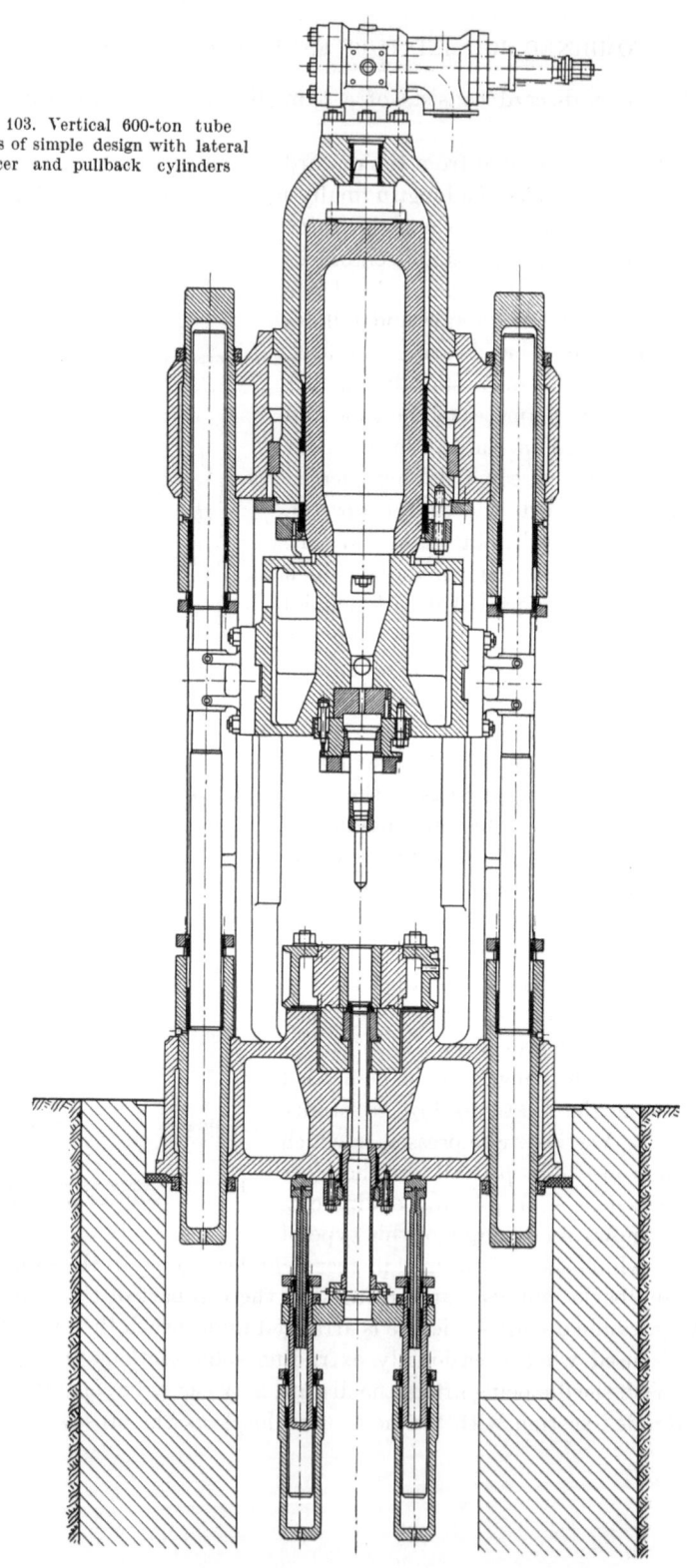

Fig. 103. Vertical 600-ton tube
press of simple design with lateral
piercer and pullback cylinders

such extrusions are always small ones. Contrary to the tube die, shown
in the sketch, the die for the extrusion of sections must not have a collar
and its outside diameter must be identical to the bore of the container.
The tubes are still separated from the discard by the shearing mandrel
which, however, is not suitable to cut off various shaped and thin solid
sections.

The die used for the extrusion of sections is lifted by means of a tube
which passes through the container support and is set on a sleeve which
is connected to the ejector crosshead. The crosshead is moved up by the
two bottom plungers being so designed to form at the same time the
cylinders for the two stationary top pullback plungers which are bolted
to the frame by two tie-rods each. In order to facilitate dismounting of
the sleeves in the downward direction, the pistons can perform an excess
stroke after the bolts, which hold the sleeve in the crosshead, have first
been released.

In Fig. 104 is illustrated the largest vertical tube extrusion press built
hitherto, designed for a total capacity of 2,500 tons and provided with a
piercer operating independently of the ram.

The shape of this press is as that of the two presses described above.
Prior to installing the movable crosshead – contrary to Fig. 103 – the
main cylinder a is inserted into the bore of the frame from below. The
plunger b is sealed by a radially adjustable stuffing box, guided in a
bronze bush, and provided at its bottom with a throttle bolt to prevent
the plunger from crashing against the cylinder bottom on which is fixed
the filler valve (cushioned cylinder). The crosshead c which moves in
long guides in the frame, is – similar to Fig. 88 and 89 – slit to accomodate
the piercer crosshead and provided at its bottom with the ram-holder d,
the ram flange e and the stroke limiting collar f. The extrusion power is
transmitted on the ram via a thrust piece in which the mandrel bar g
which is attached to the piercer crosshead, is guided. It serves to hold
the mandrel-holder h which moves in the bore of the ram, with screwed-
in piercing mandrel i.

The two piercer cylinders j are arranged at the sides of the main
cylinder. The piercer plungers k form at their bottom ends the cylinders
for two stripper plungers l, being attached to the crosshead by means of
brackets. The cylinders are fed with controlled pressure water through a
bore provided in either piercer plunger and telescopic tubes m connected
thereto, which immerge into fixed conduits connected to the bottom of
the piercer cylinders. With the ram being lowered, the piercer crosshead
is lifted and the mandrel made to retract inside the ram for the upsetting operation. In case of mandrels of varying length the faces of mandrel
and dummy-block are made flush by placing shims n on bolts on the
piercer crosshead, thus changing the relative stroke of the latter towards

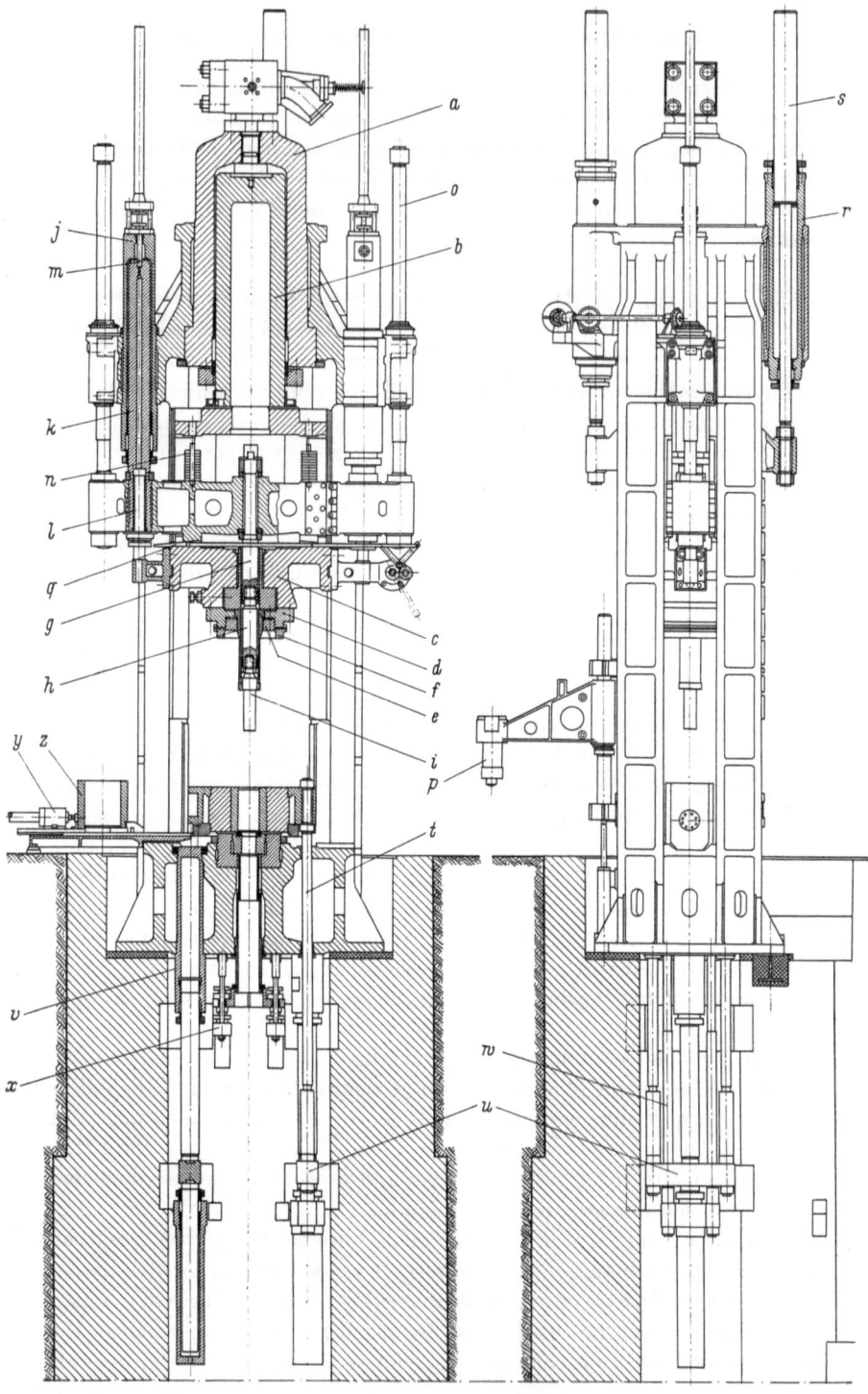

Fig. 104. Vertical 2,500-ton tube press with piercer working independently of the main ram and with movable container (By Hydraulik, Duisburg)

the top. When making tubes with the mandrel being arrested in the die, the mandrel is adjusted to a given, relative position in the die by the two external stroke limiting rods o. The latter move in threaded bushes which are attached to the frame and are uniformly adjusted on either side by a common driving shaft and form the stops for the top nuts of the rods.

By placing all of the shims on the piercer crosshead, it is also possible to extrude tubes with floating mandrel. In this case the billet cannot be upset prior to piercing with the piercing mandrel having been retracted, and the press is therefore provided with a pivoted upsetting sleeve p which embraces the mandrel i and joins in the motion of the ram to perform the upsetting operation. When upsetting the billet by this sleeve, a small hole is punched into the billet to center the mandrel.

When extruding with floating mandrel, the piercer- and main cylinders are simultaneously fed with pressure water. Piercer crosshead and main crosshead are thus coupled so that – contrary to the extrusion with fixed mandrel – both the extrusion and piercer powers operate on the ram. In order to be able to use the mandrel collar for the cutting of the tube from the discard – instead of the pivoted shearing mandrel – opposite wedges q are arranged between main- and piercer crossheads which – on completion of the extrusion and on clearing the piercer crosshead – are shifted back by hand with the help of a linkage, thus allowing the piercer crosshead to make a short excess stroke for the shearing operation (see p. 149 and Fig. 121, c).

The stripper cylinders incorporated in the piercer plungers k permit of using a common pullback device for main crosshead and piercer crosshead. The two pullback cylinders r are arranged on the front and back sides of the frame and the pullback plungers s are of the differential type. They may just as well be simple plunger pistons provided with a reversing linkage.

The container-holder with container is guided in the bottom part of the frame in the same manner as the main crosshead and may be lifted from the die. This arrangement was made to permit of cutting the tube or bar – when extruding on the "with shell" method – by a sliding saw mounted on the front side, which is not shown in the sketch. The container is lifted and lowered by four rods t which are secured in two crossheads u underneath the frame. The cylinders v of the plungers for the downward motion are arranged in the frame base which also holds – by tie-rods w – the cylinders for the upward motion. The long plunger stroke permits of easily detaching container and ram towards the bottom. A small lifting- and pullback device x is provided at the bottom edge of the frame base to allow dismounting of die and die-holder.

For the purpose of ejecting "stickers" provison has been made on the narrow side of the press for a shifting device y by means of which a

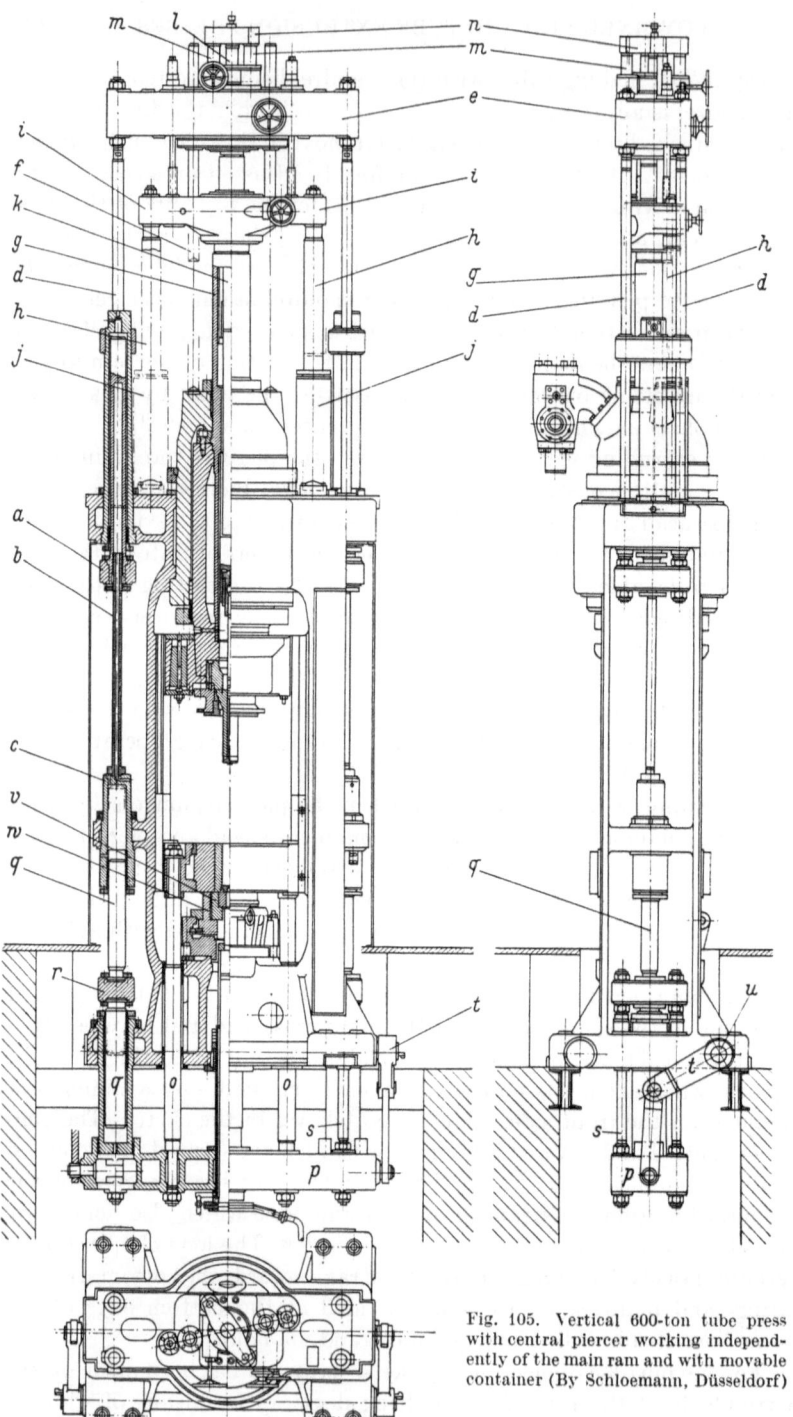

Fig. 105. Vertical 600-ton tube press with central piercer working independently of the main ram and with movable container (By Schloemann, Düsseldorf)

bush z is shifted under the lifted container to support it and to take up
the ejected billet.

Fig. 106. Shop photo of press as per Fig. 105

The design of a vertical 600-ton tube press with a central piercer bar
passing through the main plunger, is illustrated in Fig. 105 and 106. The
two piercer cylinders are inserted in the press frame at either side of the

main cylinder. The piercer plungers press on two crossheads a and serve also as cylinders for two stationary piercer pullback plungers b that are fixed on two lateral cylinders c, attached to the press frame, for the descent and the sealing of the container. The crossheads a are connected to the moving piercer crosshead e through four rods d. When extruding with fixed mandrel, the downward stroke of the moving piercer crosshead is adjustable by means of the two stroke limiting rods f. The rods are uniformly adjusted by nuts which are rotated by a hand-wheel through a gear, and rest on the cylinder bottom. The piercer bar k is reliably guided in a long extension g which is bolted to the base of the main plunger and also serves to retract the main plunger when the pullback plungers h push the crosshead i upwards. The piercer bar k consists of two parts. The bottom part which carries the mandrel-holder with the piercing mandrel, is pressed against the hollow-bored top part which is secured in the crosshead e by the anchor l, as the latter is held under a constant traction by two small plungers m through the crosshead n. When a given traction of the piercing mandrel is exceeded, the piercing mandrel can therefore give way against the pressure of the plungers m. However, such elastic holding device for the mandrel necessitates all the above details as well as movable feed pipes for the plungers m.

The container is lifted and lowered – as with the press illustrated in Fig. 105 – by means of two rods o and a common crosshead p. The crosshead is driven by plungers q with their links r and four lateral tie-rods s. In order to prevent a tilting of the crosshead in case of non-uniform plunger powers and due to the long distance between crosshead and cylinder, the crosshead is laterally connected to a through shaft u by levers t, thus providing a reliable guide.

The die-holder v is set on a rotary table w which is provided with a worm gear and may be rotated by a hand-operated worm. This device is desirable when extruding hollow sections with the help of an angular mandrel, so as to be able to set the mandrel into the proper relative position to the die. The runout system for the extrusions consists of a pipe, being arranged in the center line of the frame, with a cooling jacket.

When extruding rods and tubes on the "with shell" method, a shear – not shown in the sketch – is provided for separating the billet discard from the die.

Chapter IV

RATING AND DESIGN RULES FOR
ROD AND TUBE EXTRUSION PRESSES

a) Determination of Piston Forces

Calculation of the extrusion force, i. e. of the necessary ram force, at close approximation is only possible in few, particularly simple cases, because it depends on a variety of factors that can only be evaluated in an imperfect manner. These factors include the resistance to deformation of the alloy to be extruded, in dependence of the temperature and speed at which the work is performed, the resistances caused by the varying flow phenomena in the billet, the frictional resistance on the tools, the extrusion ratio, the shape of the section, etc. The latter factor alone may essentially increase the required force, if the section is of intrical shape as compared with round rods of identical cross-sectional area (see Fig. 107).

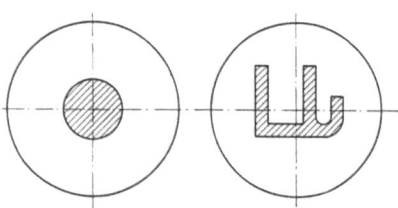

Fig. 107. Profiles of equal cross-sectional area

These calculations require first of all a clear understanding of the internal flow phenomena in the billet. An important point to be noted here is that – in direct extrusion – it is not the complete billet, as it is pushed forward by the ram, that flows towards the die, but that certain portions of the billet even flow back against the ram.

The flow of metals may easily be ascertained from experiments[1]) run on a 1,500-ton capacity rod extrusion press, in which – as illustrated in Fig. 108 – a 100 mm rod was extruded from a 163 mm container. The billet was made up with disks of 50 mm thickness and 150 mm diameter (sketch a) alternatively of copper and brass, and first of all compressed to a diameter of 163 mm and a total billet length of 592 mm (sketch b). Sketches c–e show clearly how the speed of flow is slowed down by the friction of the billet on the container wall and how the various disks bulge out towards the die aperture. From the movement of the various disks it is seen that disk no. 8, for example, is dislodged for a distance of 205 mm from the die after a ram stroke of 92 mm (see sketch c). In sketch b the distance between this very edge and the die – in the com-

---

[1]) BERNHOEFT, C.: Arbeitsverhältnisse an einer direkt angetriebenen 1500-t-Strangpresse mit 300 at Preßdruck. Z. Metallkde. 1932, H. 9.

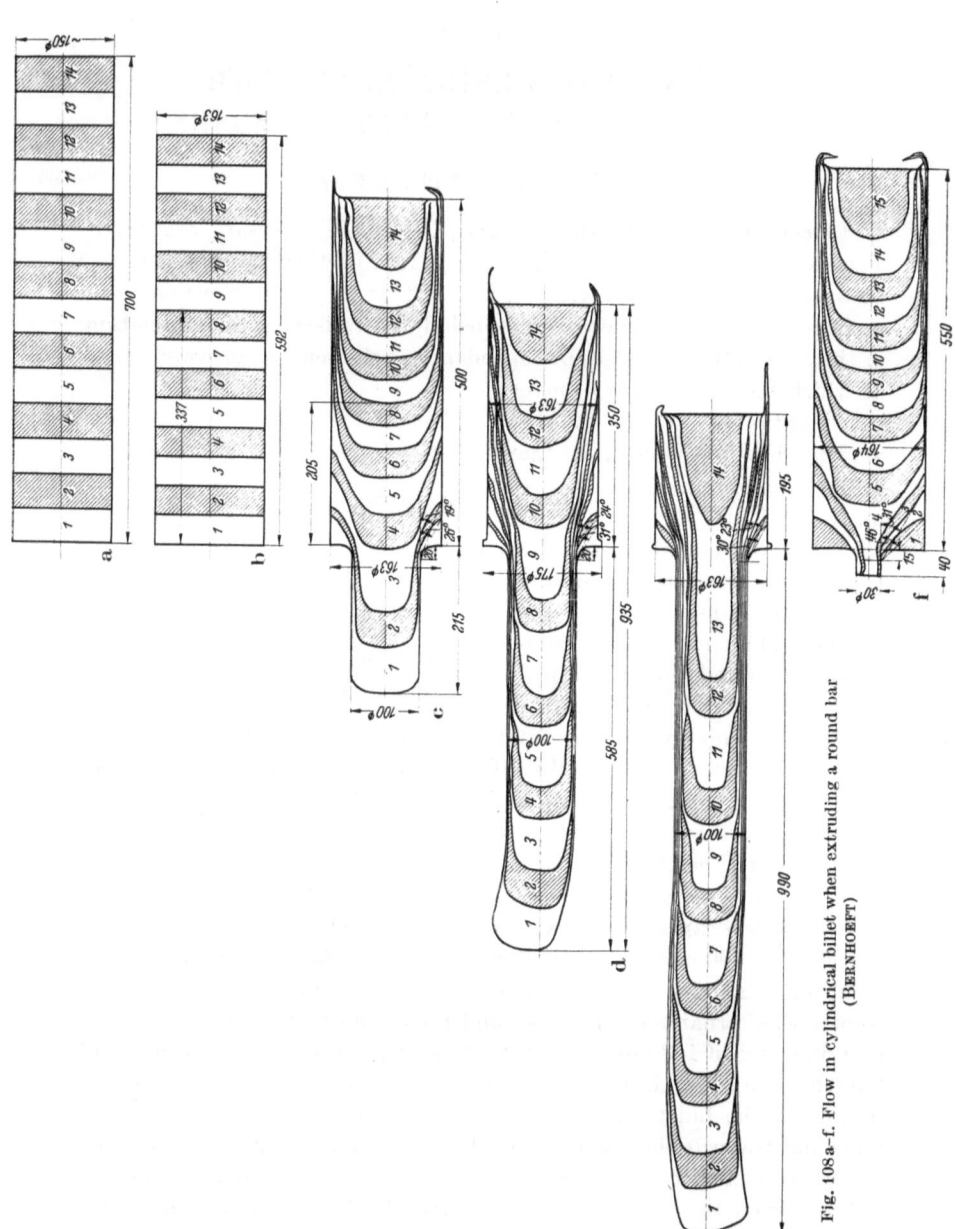

Fig. 108a–f. Flow in cylindrical billet when extruding a round bar (BERNHOEFT)

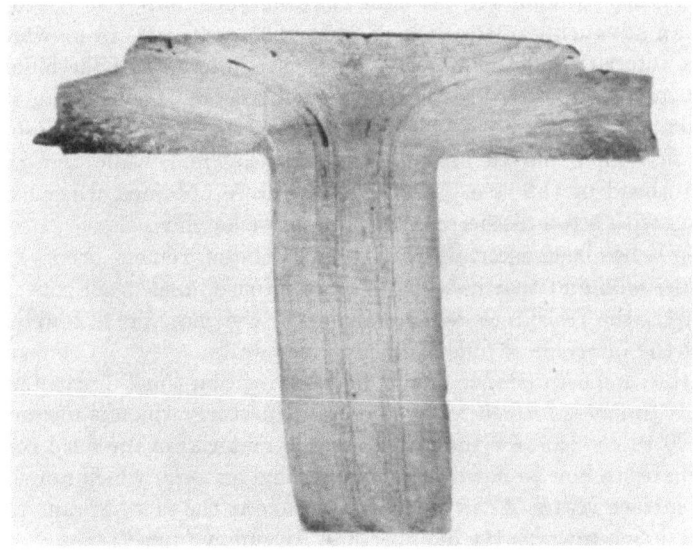

Fig. 109. Flow lines in billet residue in direct extrusion (PEARSON)

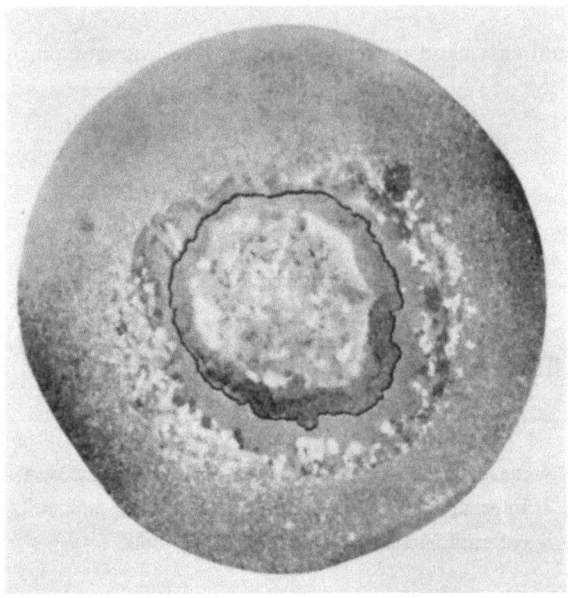

Fig. 110. Inclusion of dross in billet residue in direct extrusion (PEARSON)

pressed billet – amounts to 337 mm. The difference, $337 - 92 = 245$ mm, shows an advancing of the disk of $245 - 205 = 40$ mm, from which it may be inferred that the material at the circumference of the billet has moved in the opposite direction; in a shell zone of about 30 mm thickness particles of disk no. *8* extend as far as to the extrusion ram, so that they must have moved back $500 - 205 = 295$ mm, while the core is 40 mm ahead of the ram. Similar results were obtained when experimenting with a rod diameter of 30 mm only (sketch f).

Also it has been ascertained that only little movement occurs at the container wall and that it becomes more intense in the shell zone. Consequently, the resistance to friction inside the material is lower than that at the interface of billet and container wall.

Within a short distance of the die a "transition cone" in the form of a hollow funnel is formed towards the die aperture, which is inclined by about 20 to 45° towards the ram axis. The material in the dead corners participates in flow to hardly any extent. The material which flows back in the surface layers of the billet, is diverted at the faces of ram and/or dummy-block towards the die aperture. An inflow funnel is thus formed which – towards the end of extrusion – is liable to be occupied by fragments of dross and oxide drawn out of the skin[1] (see Fig. 109 and 110). This disadvantage is obviated by leaving part of the billet unextruded, the thickness of this discard corresponding approximately to that of the dummy-block.

The actual extrusion power is given by the known equation[2]

$$P = F_0 \ln \varphi \, k_w = F_0 \ln \varphi \, \frac{k_f}{\eta},$$

in which:

F_0 = face area of dummy-block pressing on billet,

$\varphi = \dfrac{F_0}{F_1}$ = extrusion ratio,

F_1 = cross-sectional area of extrusion,

k_w = deformation resistance of metal.

The deformation resistance is given by

$$k_w = k_f + k_i + k_r \,,$$

in which:

k_f = mean tensile strength of metal during deformation free of loss,

k_i = internal frictional resistances in the flowing material,

k_r = external frictional resistances at the tools.

---

[1] PEARSON, C. E.: The Extrusion of Metals. London: Chapman & Hall 1953.
[2] SIEBEL, E.: Die Formgebung im bildsamen Zustand. Düsseldorf: Stahleisen 1932.

The mean tensile strength k_f may be assessed by experiments; it depends on the amount of deformation, the speed at which deformation is carried out and the temperature of the material.

The frictional resistances k_i and k_r are empirical values. They are dependent on the shape of the tools, their surface finish, lubrication etc. and represent a very uncertain factor in the calculation which may greatly affect the final result.

The forming efficiency $\eta = \dfrac{k_f}{k_w}$ is the ratio of mean tensile strength v. deformation resistance and lies in general between 0.3 and 0.6.

Another theory establishes the equation[1])

$$P = P_1 + P_2 ,$$

in which P_1 is the force which, according to Fig. 111, pushes the metal in Section $I - I$ through the funnel in front of the die aperture, while P_2 is the force required to overcome the resistance to flow in the billet. If $p =$ the axial stresses within section $I - I$ and $f_1 =$ the face area of the ram, it follows

$$P_1 = p \times f_1 .$$

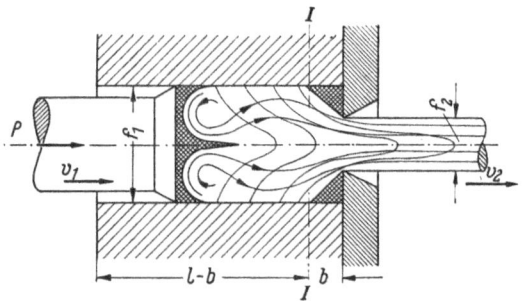

Fig. 111. Sketch explaining the determination of the extrusion force

Force P_2 is proportional to the billet volume $V = f_1 (l - b)$, the strains p within section $I - I$ and a function ψ, which depends on the extrusion speed v_1 and a constant C, so that P_2 is expressed by

$$P_2 = \psi p V .$$

Hence

$$P = p f_1 + \psi p V .$$

Using values obtained from experiments, it may be substituted

$$\psi = C \sqrt[4]{\frac{v_1}{l - b}}.$$

With normal extrusion temperaturs $C = 0.025$.

The axial stresses are given by

$$p = 2 k_m - k_f$$

[1] GELEJI, A.: Walzwerks- und Schmiedemaschinen, S. 203. Berlin: VEB Verlag Technik 1954.

in which k_m = the mean resistance to deformation, which occurs to the same extent when drawing a bar through a conical draw die.

It follows

$$k_m = \frac{k_f}{1 - 0.93 \dfrac{\varDelta_f}{f_1}},$$

in which:

k_f = mean tensile strength (yield point of material in kg/cm²),
$\varDelta_f = f_1 - f_2$ = reduction in cross-section in cm²,
f_1 = face area of ram,
f_2 = cross-section of extrusion.

The equation for k_m is based on a coefficient of friction of $\mu = 0.6$ for the internal resistances and a funnel angle of $\alpha = 45°$. Both values may vary over a wide range.

In practice this calculation is usually very much simplified by establishing the equation

$$P = F\,p\,,$$

wherein:

F = face area of dummy-block
p = extrusion pressure required, which also includes the efficiency of the press.

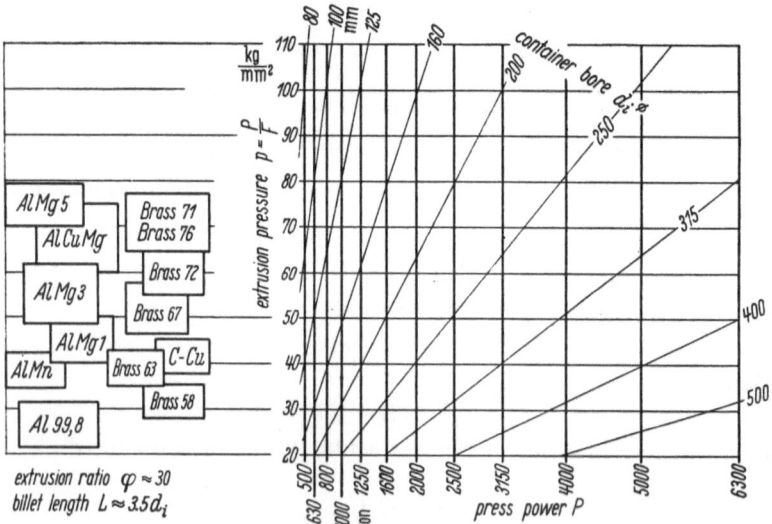

Fig. 112. Extrusion pressure p in dependence of extrusion force P and container bore d_i. Values as per standard series R 10. Properties of extrusion of various metals and alloys at ordinary temperature and speed for extrusion of round bars

The required extrusion pressure p is derived from empirical values, some of which may be taken from Table 5 and the diagram shown in Fig. 112. The values of Fig. 112 are based on a mean extrusion ratio, which

Table 5. Extrusion Properties of the Nonferrous Metals in Extrusion Presses (After K. Laue)

Nonferrous Metal Group	Alloy	Approximate Composition						Maximum Extrusion Temp. (Billet Temp.) °C	Maximum Extrusion Pressure p Required kg/mm²	Maximum Extrusion Ratio φ
		Cu	Zn	Sn	Al	Pb	Ni			
1	2	3						4	5	6
Copper and CuZn Alloys	C Cu	99.5	—	—	—	—	—	—	900	80	280
	Ms 58 [1]) and SoMs [2])	58	42	—	—	(2)	—	—	700	70	700
	Ms 63	63	37	—	—	—	—	—	750	70	600
	Ms 68	68	32	—	—	—	—	—	800	80	450
	Ms 87	87	13	—	—	—	—	—	900	80	100
Condenser Tube Alloys	70/29/1	70	29	1	—	—	—	—	900	100	80
	76/22/2	76	22	—	2	—	—	—	900	100	80
Bronzes and Difficultly Extrudable Special Bronzes	Al Bz 4	96	—	—	4	—	—	—	900	100	100
	Al Bz 9	91	—	—	9	—	—	—	920	100	100
	Sn Bz 4	96	—	4	—	—	—	—	850	100	30
	Sn Bz 8	92	—	8	—	—	—	—	750	100	30
	Pb Bz 4	96	—	—	—	4	—	—	650	100	30
	Si Bz 2	96	—	Si 3	—	—	—	Mn 1	700	100	30
	Special Bz	90	—	6	—	—	4	—	800	100	30
	Ag Bz	98	—	—	—	—	—	Ag 2	900	80	50
Copper-Nickel Alloys	Cu Ni 70/30	70	—	—	—	—	30	—	900	100	30
	Cu Ni Zn 72/18/10	72	—	10	—	—	18	—	850	100	30
	Ni 98	—	—	—	—	—	98	—	1,100	100	80
		Al	Mg	Mn	Cu	Si	Zn			
Magnesium and Mg-Alloys	pure Mg	—	99.8	—	—	—	—	300	80	200
	Mg Mn	—	R[3])	2	—	—	—	420	80	100
	Mg Al 3	3	R	—	—	—	—	400	80	80
	Mg Al 6	6	R	—	—	—	—	380	100	60
	Mg Al 9	9	R	—	—	—	—	360	100	60
Aluminum and Al-Alloys	pure Al	99.8	—	—	—	—	—	500	80	1,000
	Al Mn	R[3])	—	1.2	—	—	—	500	80	500
	Al Mg Si	R	1.2	0.8	—	1.4	—	480	80	250
	Al Mg Mn	R	2.5	1	—	—	—	450	80	100
	Al Mg 3	R	3	0.2	—	—	—	450	100	80
	Al Cu Mg	R	1.5	1	4	—	—	450	100	80
	Al Mg 5	R	5	0.5	—	—	—	420	100	70
	Al Mg 7	R	7	0.5	—	—	—	420	100	60
	Al Mg 9	R	9	0.5	—	—	—	400	100	60
	Al Zn Mg Cu	R	2.7	0.5	1	—	6	400	100	80
Zinc and Zn-Alloys	fine Zn	—	—	—	—	—	99.5	200	80	200
	Zn Al 4	4	—	—	—	—	R[3])	300	100	60
	Zn Cu 4	—	—	—	4	—	R	350	100	60
	Zn Al Cu	4	—	—	1	—	R	350	100	50

[1]) Ms = brass.   [2]) SoMs = special brass.   [3]) R = balance.

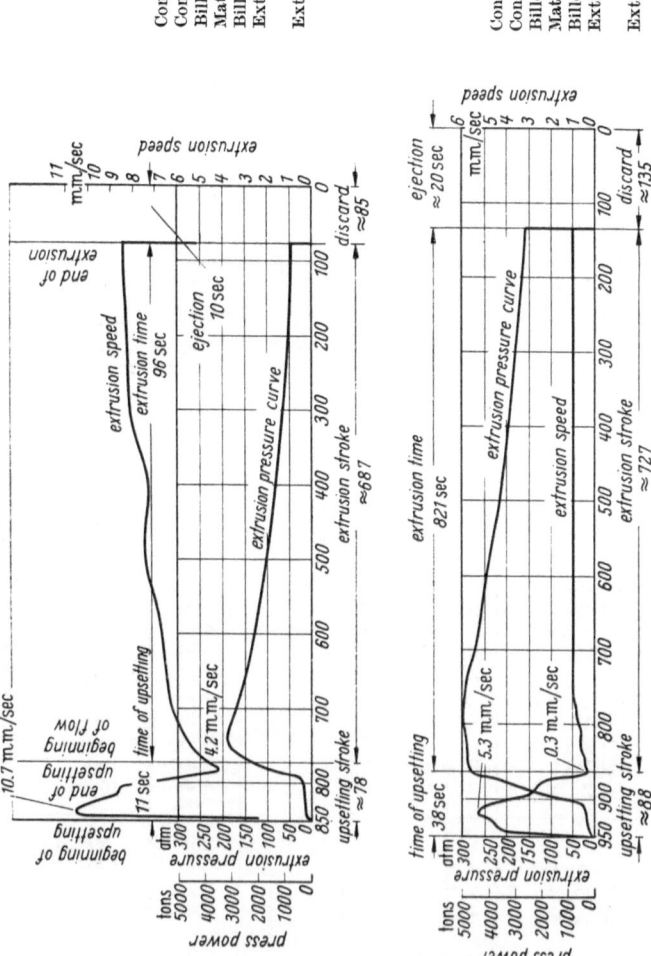

Container: 360 mm Ø.
Container temperature: 350 °C., gas heated.
Billet: 345 mm Ø × 850 mm lg.
Material: VLM 51 (initial material for forgings).
Billet temperature: 470°C.
Extruded: Round bar 250 mm Ø (cross-sectional area 490.9 cm²).
Extrusion ratio: $\varphi = 2.04$.

Container: 360 mm Ø.
Container temperature: 350 °C., gas heated.
Billet: 345 mm Ø × 950 mm lg.
Material: AlCuMg Fl. W. 3425.5.
Billet temperature: 480°C.
Extruded: 4 profiles, total cross-sectional area 87.2 cm².
Extrusion ration: $\varphi = 11.7$.

Fig. 113. Pressure and speed diagrams[1] taken on a 5,000-ton rod press when extruding aluminium alloys

[1] HEMMERICH, F.: Die Blockaufnehmer von Metallstrangpressen. Metallwirtschaft XXIII (1944) H. 27/30.

should not exceed $\varphi = 40$ to 60. On the other hand it is important that the extrusion ratio is never less than $\varphi = 10$ so as to ensure a good

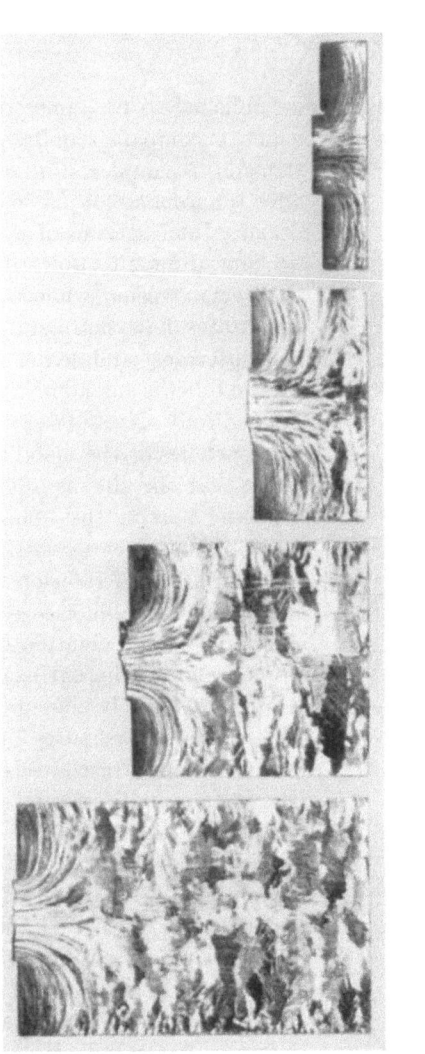
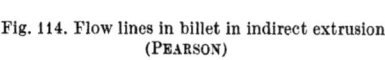
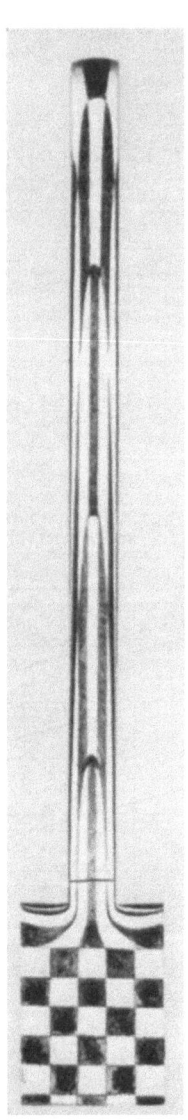

Fig. 114. Flow lines in billet in indirect extrusion (PEARSON)

Fig. 115. Flow lines in round bar in indirect extrusion (PEARSON)

deformation of the material. Special cases – such as extrusion of wire and cable sheathing – call for extrusion ratios of $\varphi = 500$ to 1,000.

The curves of extrusion pressure and extrusion speed, illustrated in Fig. 113, give a good picture of the characteristic extrusion phenomena.

Having determined the press power, the diameter of the piston or plunger D of the press is given by the equation

$$\frac{\pi D^2}{4} = \frac{P}{p_f},$$

in which p_f is the hydraulic pressure.

All the considerations and findings established so far apply to the direct method of extrusion. The flow which accompanies indirect extrusion, is simpler. Fig. 114 is the photomicrograph of the longitudinal section of a billet in four different stages during indirect extrusion, while Fig. 115 illustrates how the metal from the outermost and center zones of the billet is spread in the extruded rod[1]). It can be seen how the particles of the metal in the zone near the die go into the rod and how in the remaining part of the billet – contrary to flow in direct extrusion – no particles of the billet core move forward. The calculation of the force required in extrusion in this case may be based upon essentially lower figures for flow and frictional resistances and the saving in extrusion force averages 25 to 30% (see Fig. 116). BERNHOEFT[2]) came to the same conclusion during investigations carried out with a 1,500-ton capacity press (see Fig. 117).

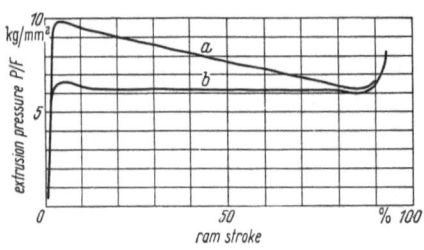

Fig. 116. Pressure curves in direct (a) and indirect (b) methods

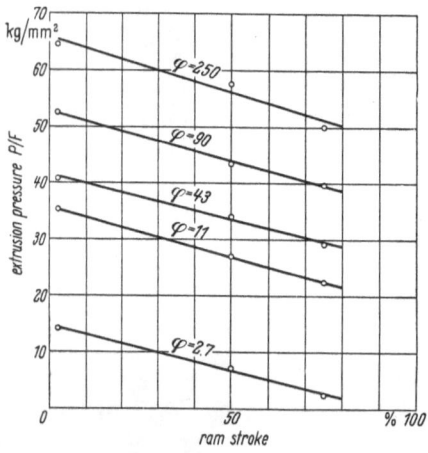

Fig. 117. Initial pressures in direct method of extrusion in dependence of the length of the billet

The indirect method has, however, not found general application on an industrial scale, because of the smaller circumcircle available for the extrusions (see Fig. 56) and because of bad access of the die. However, it may well be

[1]) PEARSON, C. E.: The Extrusion of Metals. London: Chapman & Hall 1953.

[2]) BERNHOEFT, C.: Kraftersparnis zwischen direktem und indirektem Strangpressen. Z. Metallkde. 1933 H. 12.

Fig. 118. Combined 12,000-ton tube and rod press in four-column design with 3,000-ton piercer for the extrusion of max. 800 mm diameter and 1,800 mm long light metal billets. Foreground: 1,000-ton capacity shape stretcher. Courtesy of Dow Chemical Co., Madison/USA (By: Hydraulik, Duisburg)

anticipated that the economical exploitation of the extrusion power will make this method more attractive in the future.

In most cases the pullback power chosen for the extrusion ram amounts to 10% of the extrusion power; the counterpressure in the main cylinder should not exceed about 10 to 12 atmospheres if the plunger area was designed for a hydraulic pressure of 300 to 400 atmospheres. Under these circumstances idle speeds of about 200 to 300 mm/sec are attained in either direction. With a hydraulic pressure of 200 to 300 atmospheres the permissible counterpressure is 6 to 8 atmospheres.

When extruding machined billets on the "without shell" method with the ram moving "close-fitting" in the container bore, it is recommended to increase the pullback power by 25%. If the press is equipped with advance- or piercer cylinders, the plungers of which move the main cross-head, as shown in Fig. 88, the maximum counterpressure occurring in the main cylinder is 2 to 3 atmospheres; thus the pullback power required is decreased correspondingly.

The piercing power is also rated at 10% of the extrusion power; this calculation is based upon a pressure of 30 to 40 kg/mm² on the face of the piercing mandrel. It is in special cases only that a higher piercing power is chosen when, for instance, piercing is carried out with a thick mandrel and the end of the container bore is closed by a blank die. In any case, pullback power and piercing power should be carefully verified when the maximum mandrel diameter is known[1]).

The highest pressure capacities of press installations built are about 12,000 tons with piercing powers up to about 3,000 tons (see Fig. 118).

b) Hydraulic Pressures

The hydraulic pressure to be chosen when calculating the plunger diameters of a tube or rod extrusion press depends above all on the method of operation. If the press is of the accumulator-driven type, the most advantageous operating pressure is 200 to 315 atmospheres, because in this case best use is made of the cylinder material and highest economy is ensured with regard to prime- and maintenance cost and expenses caused by the wear of control gears and packings. An extrusion power of 4,000 to 5,000 tons is considered to be the limit at an operating pressure of 200 atmospheres. For higher extrusion capacities a hydraulic pressure of 315 atmospheres is indicated.

With presses operated directly off the pumps preference is given to higher operating pressures of 400 to 500 atmospheres so as to keep the plunger diameters to a minimum. The presses concerned are mainly used

---

[1]) MÜLLER, E.: Hydraulische Pressen Bd. 2, I. Abschn.: Lochpressen. Berlin/Göttingen/Heidelberg: Springer 1955.

for rod extrusion, operating at low speeds; the maximum capacity is used temporarily only and their controls are balanced by poppet valves.

When using variable-delivery pumps for oil-hydraulic presses, the appropriate operating pressure is about 200 atmospheres as leakage losses would be excessive if this pressure will be exceeded. On the other hand, oil-driven presses with valve-controlled constant delivery pumps are – in most cases – arranged for operating pressures of 315 to 400 atmospheres.

c) Nomenclature of Tools and Tool Steels

Fig. 119 illustrates a tooling assembly for rod- and tube extrusion. Various terms have been in use for extrusion press tools which do not always refer to the same item in different plants. This has induced the

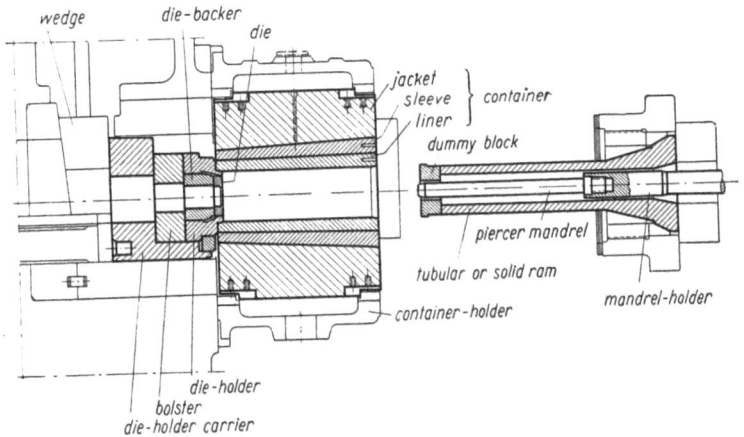

Fig. 119. Tools of a tube and rod extrusion press

standardization committee "Werkzeuge der NE-Metalltechnik", set up some years ago, to work out as a first step proposals for a uniform nomenclature of such tools[1]).

The recommendations are shown in Fig. 119. Further recommendations include the choice of materials upon which the economy of operation of the presses depends above all. Tables 6 through 9, which are derived from the material standard specifications published by *Verein Deutscher Eisenhüttenleute*, contain such details.

---

[1]) LAUE, K.: Die Benennung der Strangpreßwerkzeuge und ihre Stahlauswahl. Z. Metallkde. 1955 H. 4.

Table 6. Tool Steels for Working of Nonferrous Metals

	Designation of Tools	Used for the Extrusion of		Range of Application Dimensions Stresses
A	Liners	Light and heavy alloys	1	general
			2	high thermal and mechanical stresses
			3	for heavy and light alloys
			4	do., preferably for heavy alloys
			5	difficultly extrudable heavy alloys
			6	do., for vertical presses
B	Sleeves	Light and heavy alloys	1	general
			2	highly stressed
C	Jackets	Light and heavy alloys	1	heated container
			2	high thermal stresses
			3	do., continuous temperature > 500 °C

Table 7. Tool Steels for Working of Nonferrous Metals

	Designation of Tools	Used for the Extrusion of		Range of Application Dimensions Stresses
D	Extrusion Ram Heads	Light and heavy alloys	1	general
			2	for extrusion preferably light alloy
			3	temp. < 700 °C preferably heavy alloy
			4	extrusion temp. > 700 °C for vertical press
			5	(also compound tubes)
E	Extrusion Rams	for horizontal presses	1	general
			2	high mechanical stresses
			3	high thermal or mechanical stresses
			4	high thermal stresses
		for vertical presses	1	generally for light metal
			2	
			3	high thermal and mechanical stresses
			4	for heavy alloys without ram head
			5	maximum thermal stresses
			6	maximum thermal stresses

in Tube and Rod Extrusion Presses. Containers

Designation as per DIN 17 006	Material No.	Approximate Composition									Approx. UTS kg/mm²	
		% C	% Si	% Mn	% Cr	% Mo	% Ni	% V	% W	% Co		
45 Cr Mo V 67	2323	0.45	0.3	0.7	1.5	0.7	—	0.3	—	—	125 – 140	
× 38 Cr Mo V 51	2343	0.38	1.0	0.4	5.0	1.3	—	0.3	—	—	130 – 150	
45 Cr V Mo W 58	2603	0.45	0.6	0.4	1.5	0.5	—	0.8	0.5	—	130 – 150	
× 32 Cr Mo V 33	2365	0.32	0.3	0.3	2.8	2.8	—	0.5	—	—	130 – 150	A
× 30 W Cr V 41	2564	0.3	1.0	0.4	1.0	—	—	0.2	4.0	—	130 – 150	
× 30 W Cr V 53	2567	0.3	0.2	0.3	2.5	—	—	0.6	4.5	—	140 – 155	
40 Cr Mn Mo 7	2311	0.4	0.3	1.5	2.0	0.2	—	—	—	—	100 – 120	B
45 Cr Mo V 67	2323	0.45	0.3	0.7	1.5	0.7	—	0.3	—	—	110 – 130	
45 Cr Mn Mo 7	2311	0.4	0.3	1.5	2.0	0.2	—	—	—	—	90 – 110	
45 Cr Mo V 67	2323	0.45	0.3	0.7	1.5	0.7	—	0.3	—	—	100 – 110	C
× 38 Cr Mo V 51	2343	0.38	1.0	0.4	5.0	1.3	—	0.3	—	—	100 – 110	

in Tube and Rod Extrusion Presses. Extrusion Rams

Designation as per DIN 17 006	Material No.	Approximate Composition									Approx. UTS kg/mm²	
		% C	% Si	% Mn	% Cr	% Mo	% Ni	% V	% W	% Co		
45 Cr V Mo W 58	2603	0.45	0.6	0.4	1.5	0.5	—	0.8	0.5	—	140 – 160	
× 32 Cr Mo V 33	2365	0.32	0.3	0.3	2.8	2.8	—	0.5	—	—	140 – 160	
× 30 W Cr V 53	2567	0.3	0.2	0.3	2.5	—	—	0.6	4.5	—	140 – 160	D
× 30 W Cr V 93	2581	0.3	0.2	0.3	2.5	—	—	0.4	9.0	—	140 – 170	
× 30 W Cr Co V 93	2662	0.3	0.2	0.3	2.5	—	—	0.3	9.0	2.0	140 – 170	
56 Ni Cr Mo V 7	2714	0.55	0.3	0.7	1.0	0.5	1.7	0.1	—	—	150 – 180	
× 45 Ni Cr Mo 4	2767	0.45	0.2	0.5	1.3	0.2	4.0	—	(0.5)	—	150 – 180	
× 38 Cr Mo V 51	2343	0.38	1.0	0.4	0.5	1.3	—	0.3	(1.5)	—	150 – 180	
× 30 W Cr V 53	2567	0.3	0.2	0.3	2.5	—	—	0.6	4.5	—	140 – 160	
50 Ni Cr Mo V 7	2714	0.55	0.3	0.7	1.0	0.5	1.7	0.1	—	—	150 – 180	E
× Ni Cr Mo 4	2767	0.45	0.2	0.5	1.3	0.2	4.0	—	(0.5)	—	150 – 180	
× 32 Cr Mo V 51	2343	0.38	1.0	0.4	0.5	1.3	—	0.3	(1.5)	—	150 – 180	
× 30 W Cr V 53	2567	0.3	0.2	0.3	2.5	—	—	0.6	4.5	—	150 – 170	
× 30 W Cr V 93	2581	0.3	0.2	0.3	2.5	—	—	0.4	9.0	—	150 – 170	
× 30 W Cr Co V 93	2662	0.3	0.2	0.3	2.5	—	—	0.3	9.0	2.0	150 – 170	

Table 8. Tool Steels for Working of Nonferrous Metals in Tube and Rod Extrusion

Designation of Tools	Used for the Extrusion of		Range of Application Dimensions Stresses
F Extrusion Dies	Zinc and Lead Alloys	1	for tubes and rods
		2	for sections
	Light Alloys	1	rods, sections, tubes – normal stresses
		2	special sections and tubes
		3	do., high stresses
		4	do.
	Heavy Alloys	1	rods and sections – normal stresses
		2	do.
		3	sections and tubes – high stresses
		4	do., maximum stresses
		5	tubes and wire (austenitic)
		6	do., bushings (hard alloy)
	for Chamber and Bridge Dies (also Cables)	1	for zinc, lead, aluminum
		2	} do. also for section dies
		3	
		4	for high thermal stresses
G Extrusion Mandrels (Centering Mandrels)	Zinc and Lead Alloys	1	mandrels >40 mm dia.
		2	small mandrels, highly stressed
	Light Alloys	1	large tubes 50 to 100 mm dia.
		2	do.
		3	mandrels <50 mm dia
		4	do.
		5	highly stressed mandrels <40 mm dia.
	Heavy and Special Alloys	1	Extrusion temperature $<700\ °C$ water cooling
		2	do. $<800\ °C$
		3	do. mandrels >40 mm dia.
		4	mandrels >40 mm dia. oil cooling
		5	mandrels <40 mm dia. oil cooling
		6	} highly stressed mandrels, especially for vertical presses. Extr. temp. $>800\ °C <40$ mm dia.
		7	
H Dummy-Blocks	Zinc and Lead Alloys	1	general
	Light Alloys	1	general
		2	higher stresses
		3	special cases: slight tendency of adherence
	Heavy Alloys	1	normal stresses
		2	do., water cooling
		3	highly stressed. Water cooling
		4	highly stressed. Oil cooling
		5	maximum stresses

Presses. Wear Tools: Extrusion dies, extrusion mandrels and dummy-blocks

Designation as per DIN 17 200	Material No.	Approximate Composition									Approx. UTS kg/mm²
		% C	% Si	% Mn	% Cr	% Mo	% Ni	% V	% W	% Co	
61 Cr Si V 5	2243	0.61	0.9	0.8	1.2	—	—	0.1	—	—	140 – 160
45 W Cr V 77	2547	0.45	1.0	0.3	1.7	—	—	0.2	2.0	—	140 – 160
45 Cr V Mo W 58	2603	0.45	0.6	0.4	1.5	0.5	—	0.8	0.5	—	140 – 160
× 38 Cr Mo V 51	2343	0.38	1.0	0.4	5.0	1.3	—	0.3	—	—	140 – 160
× 32 Cr Mo V 33	2365	0.32	0.3	0.3	2.8	2.8	—	0.5	—	—	130 – 150
× 30 W Cr V 53	2567	0.3	0.2	0.3	2.5	—	—	0.6	4.5	—	140 – 160
× 32 Cr Mo V 33	2365	0.32	0.3	0.3	2.8	2.8	—	0.5	—	—	140 – 160
× 30 W Cr V 53	2567	0.3	0.2	0.3	2.5	—	—	0.6	4.5	—	140 – 160
× 30 W Cr V 93	2581	0.3	0.2	0.3	2.5	—	—	0.4	9.0	—	140 – 160
× 30 W Cr Co V 93	2662	0.3	0.2	0.3	2.5	—	—	0.3	9.0	2.0	140 – 160
× 50 Ni Cr V W 1313	2731	0.5	1.3	0.7	13.0	—	13.0	0.5	2.5	—	110 – 125
G × 170 Co Cr W 3325	9850	1.7	2.45	0.5	25.0	—	—	—	6.0	33.0	—
× 30 W Cr V 53	2567	0.3	0.2	0.3	2.5	—	—	0.6	4.5	—	135 – 150
(× 45 Ni Cr Mo 4)	(2767)	0.45	0.2	0.5	1.3	0.75	4.0	0.2	—	—	135 – 150
× 32 Cr Mo V 33	2365	0.32	0.3	0.3	2.8	2.8	—	0.5	—	—	135 – 150
× 30 W Cr V 93	2581	0.3	0.2	0.3	2.5	—	—	0.4	9.0	—	135 – 150
45 Cr Mo V 67	2323	0.45	0.3	0.7	1.5	0.7	—	0.3	—	—	130 – 150
× 30 W Cr V 53	2567	0.3	0.2	0.3	2.5	—	—	0.6	4.5	—	150 – 160
45 W Cr V 77	2547	0.45	1.0	0.3	1.7	—	—	0.2	2.0	—	150 – 170
45 Cr Mo V 67	2323	0.45	0.3	0.7	1.5	0.7	—	0.3	—	—	150 – 170
× 38 Cr Mo V 51	2343	0.38	1.0	0.4	5.0	1.3	—	0.3	—	—	150 – 170
× 32 Cr Mo V 33	2365	0.32	0.3	0.3	2.8	2.8	—	0.5	—	—	150 – 170
× 30 W Cr V 53	2567	0.3	0.2	0.3	2.5	—	—	0.6	4.5	—	160 – 170
× 38 Cr Mo V 51	2343	0.38	1.0	0.4	5.0	1.3	—	0.3	—	—	130 – 150
× 32 Cr Mo V 33	2365	0.32	0.3	0.3	2.8	2.8	—	0.5	—	—	140 – 150
× 30 W Cr V 41	2564	0.3	1.0	0.4	1.0	—	—	0.2	4.0	—	140 – 150
× 30 W Cr V 53	2567	0.3	0.2	0.3	2.5	—	—	0.6	4.5	—	150 – 160
× 30 W Cr V 93	2581	0.3	0.2	0.3	2.5	—	—	0.4	9.0	—	160 – 170
× 30 W Cr Co V 93	2662	0.3	0.2	0.3	2.5	—	—	0.3	9.0	2.0	160 – 170
× 65 W Cr Mo V 94	2584	0.65	0.3	0.3	4.0	0.9	—	0.7	9.0	—	160 – 170
40 Cr Mn Mo 7	2311	0.4	0.3	1.5	2.0	0.2	—	—	—	—	130 – 150
45 Cr V Mo W 58	2603	0.45	0.6	0.4	1.5	0.5	—	0.8	0.5	—	140 – 160
45 W Cr V 7	2542	0.45	1.0	0.3	1.1	—	—	0.2	2.0	—	140 – 160
× 38 Cr Mo V 51	2343	0.38	1.0	0.4	5.0	1.3	—	0.3	—	—	140 – 160
45 Cr Mo W 58	2603	0.45	0.6	0.4	1.5	0.5	—	0.8	0.5	—	130 – 150
× 30 W Cr V 41	2564	0.3	1.0	0.4	1.0	—	—	0.2	4.0	—	130 – 150
× 32 Cr Mo V 53	2365	0.32	0.3	0.3	2.8	2.8	—	0.5	—	—	140 – 160
× 30 W Cr V 53	2567	0.3	0.2	0.3	2.5	—	—	0.6	4.5	—	140 – 160
× 30 W Cr V 93	2581	0.3	0.2	0.3	2.5	—	—	0.4	9.0	—	140 – 160

The right margin is grouped by brackets labelled **F** (rows 1–16), **G** (rows 17–30), and **H** (rows 31–40).

Table 9. *Working of Nonferrous Metals in Tube*

	Designation of Tools	Used for the Extrusion of		Range of Application Dimensions Stresses
J	Die-Holder Carrier and Die-Holder	Light and Heavy Alloys	1	general
			2	high thermal stresses
			3	high mechanical stresses
			4	for heavy alloys and on older vertical presses
			5	for maximum thermal stresses
K	Die Backers	do.	1	general
			2	do.
			3	high mechanical stresses
			4	high mechanical and thermal stresses
L	Bolsters	do.	1	general
			2 – 3	} depending on mechanical and thermal stresses
M	Tool-Holders	do.	1	general
			2	with older presses: tool-holder = die-holder
N	Extrusion Mandrel Holder	do.	1	general
			2	for vertical presses
O	Upsetting Rams, Shearing Rams, Shearing Mandrels	do.	1	general
			2	higher mechanical stresses
P	Ejector Disks	do.	1	general

and Rod Extrusion Presses. Auxiliary Tools

Designation as per DIN 17006	Material No.	Approximate Composition									Approx. UTS kg/mm²	
		% C	% Si	% Mn	% Cr	% Mo	% Ni	% V	% W	% Co		
40 Cr Mn Mo 7	2311	0.4	0.3	1.5	2.0	0.2	—	—	—	—	110 — 130	
45 Cr Mo V 67	2323	0.45	0.3	0.7	1.5	0.7	—	0.3	—	—	120 — 140	
56 Ni Cr Mo V 7	2714	0.55	0.3	0.7	1.0	0.5	1.7	0.1	—	—	130 — 150	J
× 30 W Cr V 53	2567	0.3	0.2	0.3	2.5	—	—	0.6	4.5	—	110 — 130	
× 30 W Cr V 93	2581	0.3	0.2	0.3	2.5	—	—	0.4	9.0	—	120 — 140	
45 Cr Mo 67	2323	0.45	0.3	0.7	1.5	0.7	—	0.3	—	—	110 — 130	
45 W Cr V 77	2547	0.45	1.0	0.3	1.7	—	—	0.2	2.0	—	110 — 130	K
× 45 Ni Cr Mo 4	2767	0.45	0.2	0.5	1.3	0.2	4.0	—	(0.5)	—	120 — 150	
56 Ni Cr Mo V 7	2714	0.55	0.3	0.7	1.0	0.5	1.7	0.1	—	—	120 — 150	
40 Cr Mn Mo 7	2311	0.4	0.3	1.5	2.0	0.2	—	—	—	—	110 — 130	
55 Ni Cr Mo V 6	2713	0.55	0.3	0.6	0.7	0.2	1.7	0.1	—	—	120 — 150	L
56 Ni Cr Mo V 7	2714	0.55	0.3	0.7	1.0	0.5	1.7	0.1	—	—	120 — 150	
60 Mn Si 4	2826	0.6	1.0	1.0	—	—	—	—	—	—	80 — 95	M
× 45 Ni Cr Mo 4	2767	0.45	0.2	0.5	1.3	0.2	4.0	—	(0.5)	—	130 — 150	
55 Ni Cr Mo V 6	2713	0.55	0.3	0.6	0.7	0.2	1.7	0.1	—	—	115 — 130	N
× 45 Ni Cr Mo 4	2767	0.45	0.2	0.5	1.3	0.2	4.0	—	(0.5)	—	120 — 150	
45 Cr V Mo W 58	2603	0.45	0.6	0.4	1.5	0.5	—	0 8	0.5	—	130 — 140	O
× 38 Cr MoV 51	2343	0.38	1.0	0.4	5.0	1.3	—	0.3	—	—	140 — 160	
60 Mn Si 4	2826	0.6	1.0	1.0	—	—	—	—	—	—	110 — 130	P

d) Upsetting of Billet

In order to facilitate insertion of billets in the container, the former are made 5 to 10 mm smaller in diameter than the container bore. A sickle-shaped slit is thus formed between billet and container wall (see also Fig. 120). If the billet were pierced in this state and then extruded, the mandrel would be subjected to an up-acting pressure, which would cause the mandrel to deflect and result in tubes of non-uniform wall thickness.

Prior to piercing, the billet must therefore be upset so that it fills the container space completely; in order to do that the face of the dummy-

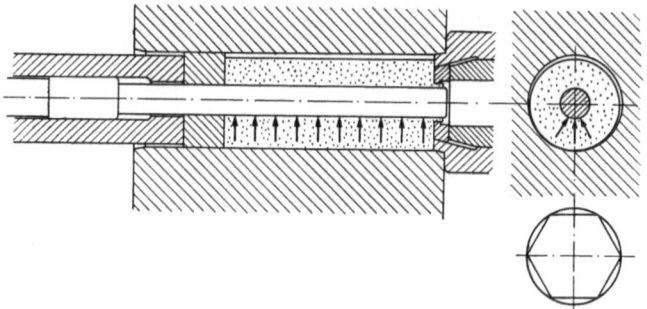

Fig. 120. Loading of mandrel when extruding a billet that has not been upset

block must be held flush with that of the mandrel. As it is impossible for the operator to control the travel of ram and mandrel with such high accuracy the mandrel bar or the piercer crosshead respectively, is dragged along by the main plunger.

The upsetting operation is of short duration and has to be performed carefully so as to prevent the billet from being forced through the die aperture. The upsetting time may be saved by employing, for example, billets of hexagonal cross-section, as shown in Fig. 120, instead of round billets. The extrusion of such billets is particularly recommended for rod extrusion presses where the mandrel is rigidly inserted in the ram. When extruding round billets, a sleeve (see Fig. 102) is employed for the upsetting of the billets, which slightly prolongs the cycle.

e) Piercing and Extrusion of Billet

Fig. 121 illustrates the tools used in various methods of tube extrusion, when – as is general practice – round billets are used. Sketch a shows the tools most frequently used when extruding tubes employing a normal mandrel, i. e. not a stepped one.

The billet having been upset, it is pierced by the mandrel. When doing this, the ram is slightly withdrawn to allow the billet to elongate. On about the last third of its stroke, the mandrel pushes a wad of metal

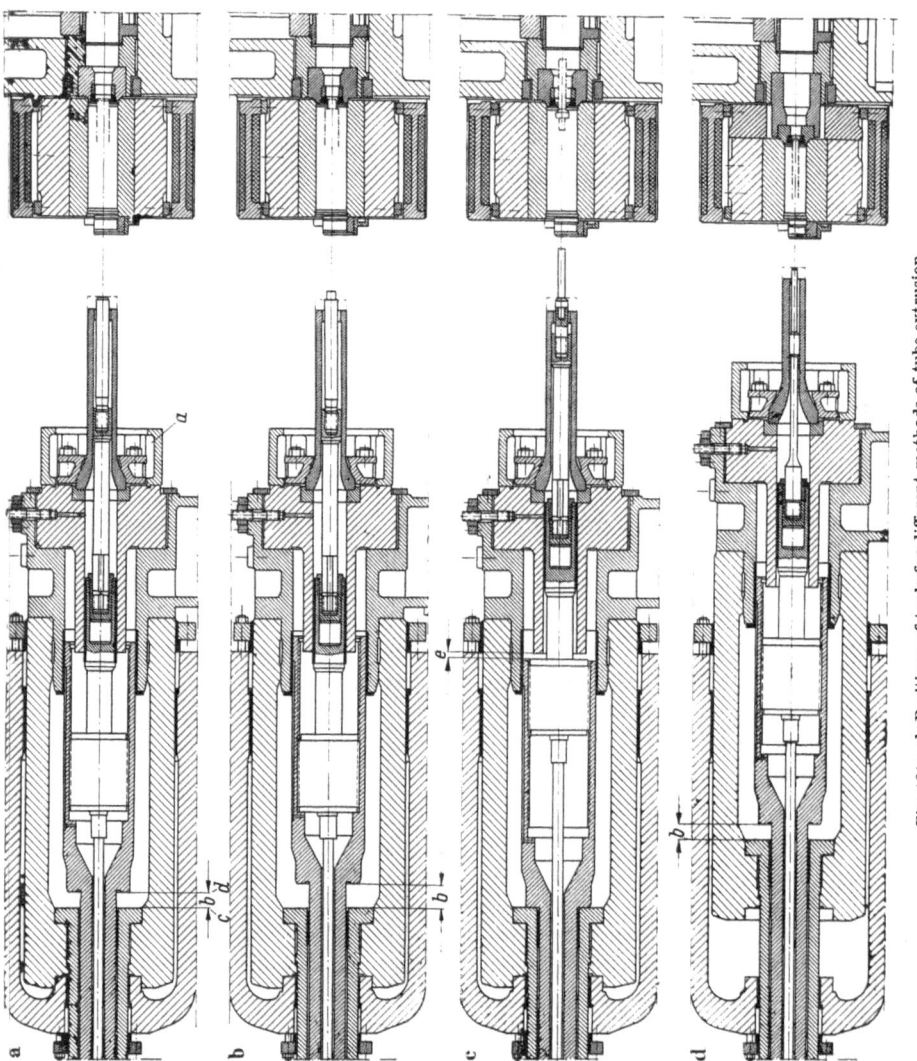

Fig. 121a–d. Positions of tools for different methods of tube extrusion

out of the billet. This wad is then removed out of the die-carrier with tongs. The piercing stroke having been completed, the piercer crosshead comes to rest against stops (see Fig. 88), when the mandrel projects about 20 to 30 mm in the die. This position is maintained when the billet is

10*

subsequently extruded by the ram through the annulus formed by die and mandrel to form the tube. The ram advances a given distance until the billet discard has attained a given thickness, which in presses of medium capacity amounts to about 50 to 100 mm. Now discard and dummy-block are ejected from the container and the die-carrier is traveled out, after having shortly relieved the main cylinder, cleared the container from the die and unlocked the die-carrier. The die-carrier may only be traveled out after the dummy-block has been ejected, as the die would be pulled off the discard and shearing would be impossible.

When ejecting discard and dummy-block, the ram stroke is limited by sleeve a coming to rest against the face of the container. This ejecting stroke must be shorter than distance b between the shoulder of the piercer bar and its stop on the plunger extension; during the idle stroke of the main plunger and during upsetting, it is maintained by two stripper cylinders being under constant hydraulic pressure (see Fig. 89). It must be noted here that the power of the stripper cylinders must be greater than the upsetting power exerted on the piercing mandrel. If this distance b did not exist and surfaces c and d touched each other, the piercer bar, which rests against the stops, would be torn off due to the extrusion power.

The effect of such stripper cylinders can also be attained by eliminating the distance between the surfaces c and d and providing a flexible stroke limiting device on the columns of the piercer device, the power of which must, however, be greater than that of the piercer, or by arranging the stops in such a way as to provide for their respective position to be adjusted or to be changed by means of spacers quickly.

If thick-walled tubes with small inside diameters, such as hollow copper bars for stud bolts, are to be extruded from the same container, tools with a pointed mandrel, as per sketch b, are employed. The operation is the same as that described above. The stripper stroke is somewhat longer, because the advance motion of the mandrel in this case is shorter than that in sketch a by the length of the mandrel point projecting over the dummy-block.

In order to obtain accurate wall thicknesses and clean surfaces when making tubes from difficultly extrudable alloys, it is appropriate to use short billets whose surfaces are machined and which have been provided with a central hole. Extrusion is performed with floating mandrel to ensure a smooth inside of the tube; furthermore the mandrel is changed after each extrusion cycle. Upsetting is not necessary, as the billet lies concentrically in the container bore after the mandrel has been inserted. In this case extrusion is carried out on the "without shell" method.

Tools used for this operation are illustrated in sketch c. Extrusion power and piercing power are coupled, the stroke limiting stops for the piercer bar removed, and the stripper cylinders disengaged; surfaces c and d touch each other. The adjusting bush in the piercer bar sleeve is advanced so far that there remains between ram and mandrel a short relative stroke e only. This stroke is employed – after completion of the extrusion – to push the mandrel, the shoulder diameter of which is equal to the outside diameter of the tube, through the die, thus eliminating shearing or sawing of the tube. The mandrel is set in the mandrel-holder with a tenon and is held in place during extrusion by the back-pressure acting on the shoulder area. The die-carrier having been traveled out, the mandrel is pulled out of the extruded tube, cooled and cleaned and is then ready to be inserted in the mandrel-holder again.

Alloys of low plasticity are often extruded under pressures which are in the vicinity of the permissible hot-strength of the tool steels. In such case cross-section and length of the billet are considerably smaller than when extruding normal alloys. Sketch d shows the arrangement of the tools for this operation. The short container is provided at its rear end with a thrust collar and set in the original holder. The die-carrier has a sleeve-like extension to accomodate the die-holder and the die. Ram, mandrel-holder and mandrel are shaped to suit the given conditions. The mandrel-bar being adjustable, shortest mandrel lengths may be used. The extrusion is performed as described for sketch a.

f) Dies and Die-Holders

The die aperture is given by the profile of the extruded rod or tube. The die consists of a cylindrical disk made of steel having a high degree of hot strength, and takes some of the highest wear which depends in the main on the billet temperature, the extrusion pressure, and the exit speed of the extruded bar.

The shape of the profile is worked out of the die body by boring, sawing, milling, etc. and then corrected and polished by hand. With austenitic material or hard alloys the profile may be made by the spark erosion method

Locally varying resistances to flow in complex sections with rims, necks, corners, and the like, are controlled by varying of the die lead-in, the bearing length, etc., so as to prevent the acceleration of flow at some points. Uneven flow of metal through the die leads to buckling and twisting of the extruded section.

The early rod presses were used for the extrusion of heavy metals only. Due to the high extrusion temperatures the dies had to be changed very often. In many cases dies, as shown in Fig. 122, are therefore employed which are inserted from the front into the die-holder with a 2 to 4° cone; they are ejected with a bolt. The cone has various disadvantages:

After a longer period of use the die-holder widens; the seating becomes loose; and the die may drop into the container when the die-carrier is traveled in, or it may damage the shearing blades when shearing off the discards. The central position is furthermore affected which leads to in-

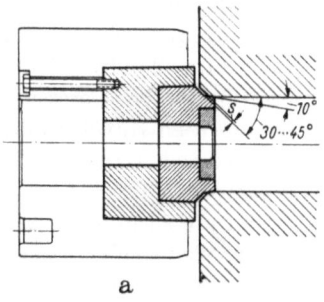

accurate wall thicknesses when extruding tubes. It is also very difficult to provide spare dies with the proper cone. If the outside diameter is for example too large, the die is not supported in the bottom of the die-holder, which may cause the die to crack during extrusion.

The die is often provided with a bayonet union to secure it; in such case the loose fit and the lacking protection against rotation give rise to objections. This type of securing the die is therefore not suitable for general application.

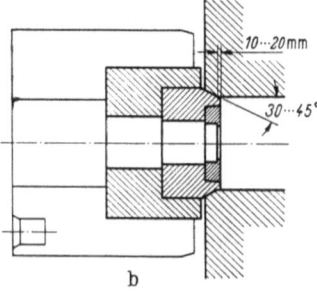

With another design, illustrated in Fig. 123, the die is arranged in front of the die-holder which is provided with a centering pin on which the die is hung. It is set loosely in the container and is thus easily exchanged. The only drawback is that the die is of large diameter and that it receives a shell after each extrusion. For cutting off the discard – when extruding with this type of die – an opposed-blade shear must be used, because the centering pin is liable to be bent when cutting in one way only.

The above drawbacks are eliminated in the design, illustrated in Fig. 124. The tapered die is inserted in the die-holder from the rear. The die-holder is easily removed towards the top after a flange has been released. The die cannot drop out of the bore in the die-holder; central

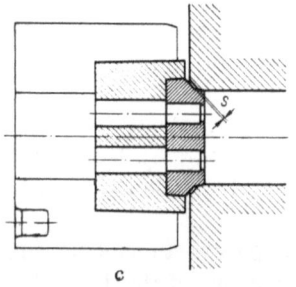

Fig. 122a–c. Arrangement of die in closed die-carrier

seating and correct support of the bearing faces are always ensured. This design is especially suitable for the extrusion of light metal billets when the die will not be changed so frequently.

The chief requisite in the production of concentric tubes is that the axes of die, mandrel and container bore coincide. Thus the die and the die-holder respectively, must be accurately centered in the container bore, while the container must be in exact axial alignment with the mandrel. Centering of the die is also to provide a sealing means to prevent metal from emerging on the circumference of the container bore.

Fig. 122 and 124 a show some examples of centering the die-holder in the container bore. Choice of one method or the other depends on the metal to be extruded and on the sealing power between container and die-holder.

Fig. 124 a illustrates the centering of the die-holder by means of a conical projection of 10° inclination and 30 mm bearing length in a light

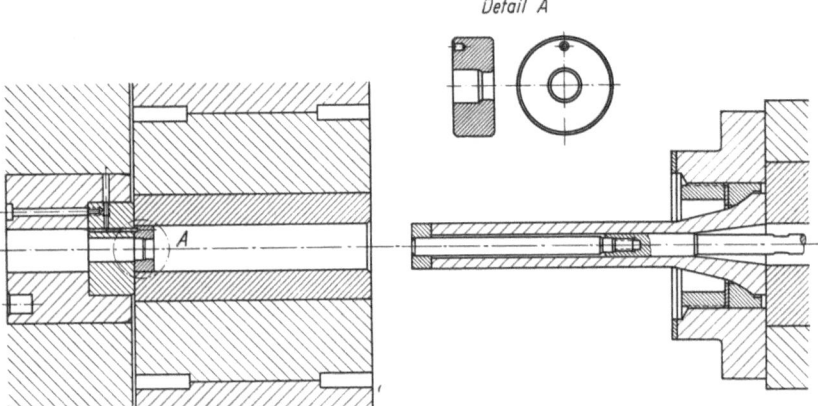

Fig. 123. Pin hung die in front of die-carrier

metal extrusion press. Excellent results have been obtained in heavy metal presses by the use of a double centering cone, as shown in Fig. 122 a and c, the inclination of which amounts to 10° and 30 to 45°. In either case the sealing power is produced by a wedge-type locking mechanism which is described in the chapter dealing with the presses. When employing flat sliding bars in connection with container shifting pistons, which ensure a more effective sealing power, resort is mostly made to a centering cone, as per Fig. 122 b, being under an angle of 30 to 45° and having a front cylindrical projection of 10 to 20 mm length. When cone centering is employed, care must be taken that the container will not butt against points other than the cone. A clearance $s = 1$ to 2 mm is therefore provided.

For the production of solid sections on rod and tube extrusion presses centering of the die in the container bore is not necessary. In order to simplify the design of the dies and to keep down the cost, experiments have been made with surface seals which, however, have proved to be successful only when extruding light metals. Fig. 124 b illustrates surface sealing for the die-holder. The circular bearing surface on the container is to be dimensioned so that the specific sealing pressure is about 10% higher than the specific ram pressure. Surface sealing is of special

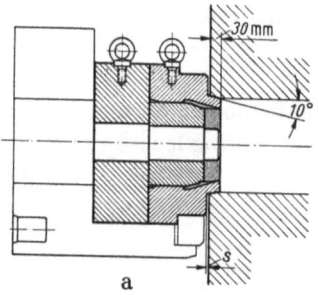

advantage in the production of bulky sections or strips, as it permits of making use of the full diameter of the container bore (see Fig. 124, c). If on the other hand it is necessary to provide the die with a conical or cylindrical sealing shoulder, the diameter of the circumcircle for the largest section must be kept essentially smaller than that of the container bore, with due regard to the remaining wall thickness (see Fig. 122, c).

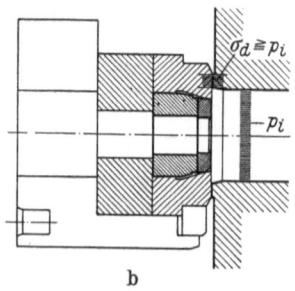

In dies for asymmetrical sections the center of gravity of the cross-sectional area of the shape to be extruded is preferably placed in the center of the die in order to obtain a satisfactory flow through the die. If it is found that the circumcircle is too small towards one side, the flow is balanced by simultaneously extruding one or more round bars (see Fig. 136). In many cases the simultaneous extrusion of blank bars can be avoided by arranging two or more asymmetrical sections opposite to one another so that the common center of gravity is again in the center of the die. When extruding round- and polygonal bars or wire, multiple-hole dies are used so as to increase the press production.

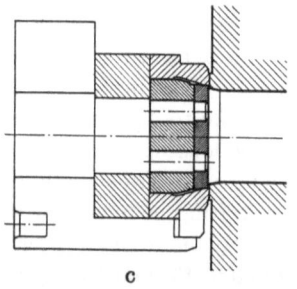

Fig. 124a–c. Arrangement of die in open die-carrier

A very advantageous characteristic of some light metals is their ability to fuse well under pressure at the extrusion temperature, this being made use of in extruding tubes and hollow sections with the help of so-called bridge- or chamber dies. Fig. 125 illustrates the design of a chamber die for a 2,500-ton tube extrusion press. It is made in two parts and consists of a mandrel-holder with die plate placed on it. The mandrel-holder is provided with a ring-shaped fusion chamber into which the metal enters through four bores of 25 mm diameter. At the die face the holes taper out to slots, so as to reduce resistance to flow. Fig. 126 shows a bridge die for hollow sections. In this case the mandrel is set on a yoke which is fitted on the die by means of two fitting pins. Extrusion with bridge- or chamber dies offers the enormous advantage that highly complex sections can be made with relatively cheap tools. Apart from this, wall thicknesses are more accurately maintained than with long mandrels.

The fusion of metal under pressure furthermore permits the extrusion of subsequent billets (billet-upon-billet) to obtain endless rods and

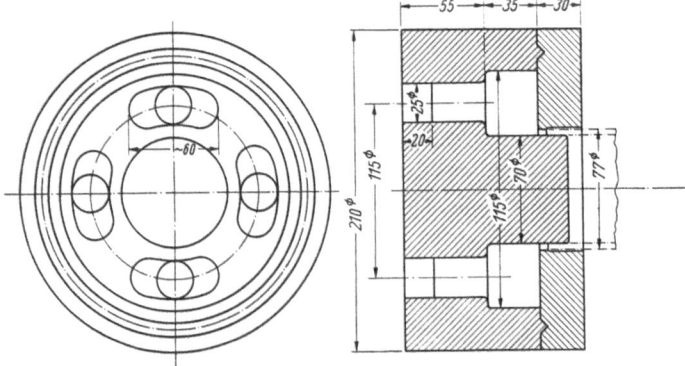

Fig. 125. Chamber die for light metals

tubes. In this case care must be taken that the billet surfaces are clean and free of oxides and that the container is well ventilated when charging a fresh billet.

As a rule the planes, in which the faces of the die and the die-holder lie, coincide and are perpendicular to the axis. Radiusing of the lead-in shoulders depends in most cases on the metal to be extruded. For aluminum and some of the aluminum alloys a nearly sharp-edged shape with a radius of 0.5 to 1 mm has given best results. With heavy metals wider radii are used. For copper and brass it is 2 to 5 mm, for cupro-nickel alloys 4 to 8 mm, and for monel 10 to 15 mm with a lead-in width of 5 to 8 mm[1]). The high radiusing is advantageous in that the die-edge is not deformed at high billet temperatures and the gauge accuracy of the section is maintained. In this case, however, the disadvantage of the greater resistance to flow must be accepted, which according to results obtained from experiments, may be about 25% higher than that in sharp-edged dies. Comparative figures are shown in Table 10.

---

[1]) BERNHOEFT, C.: Arbeitsverhältnisse an einer direkt angetriebenen 1500-t-Strangpresse mit 300 at Preßdruck. Z. Metallkde., Sept. 1932, H. 9.

Table 10. Power input for various dies on a 1,500-ton rod extrusion press with a 165 mm dia. container (After BERNHOEFT)

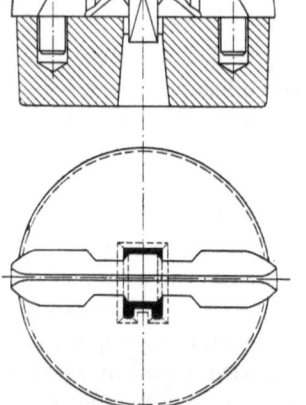

Diameter of Rod mm	Extrusion Ratio F/f	Cylinder Pressures (atm) with					
		sharp-edged die			die with lead-in		
50	11	90	50	40	110	110	60
25	43	120	100	90	130	120	110
10	268	160	150	130	220	170	160
6	750	250	280	190	290	250	240

Good results have also been obtained with dies having a lead-in tapered to 4°, as per Fig. 127.

For dies that are subject to high thermal stresses and long extrusion cycles, e. g. when extruding wire or small sections, hard-alloy bushes have proved to be very useful. They are shrunk into the dies and therefore tend less to cracking.

The endeavors to standardize the principal dimensions of dies and die-bushes have led to the proposals as outlined in Tables 11 and 12.

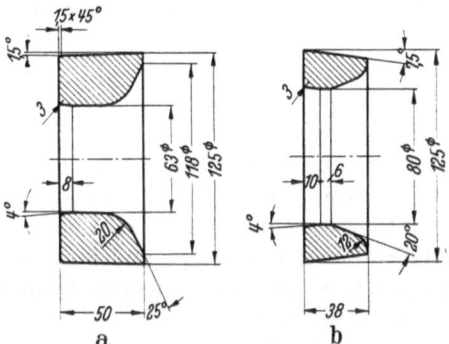

Fig. 126. Bridge die for light metals   Fig. 127a and b. Dies with tapered lead-in

A study of the patent literature reveals that many inventions refer to the working of light alloys with low plasticity which have to be extruded at very slow rates to prevent flaking and blistering of the surface of the extrusion. These are caused by the development of excessive heat and the temperature rise at the edge of the die due to excessive extrusion

Table 11. Dimensions for Dies

Diameter		Thickness h in mm		
D mm	d mm	Series 1	Series 2	Series 3
80	50	20	25	35
90	56	20	25	35
100	63	25	30	40
112	71	25	30	40
125	80	25	35	45
140	90	25	35	45
160	100	30	40	50
180	112	30	40	50
200	125	35	45	55
224	140	35	45	55
250	160	40	50	60
280	180	40	50	60
315	200	45	55	70
355	225	45	55	70
400	250	50	60	80

Table 12. Dimensions for hard alloy die bushes

Diameter		Thickness h in mm			Min. dimension of bush
D mm	d max. mm	Series 1 $h = 0.4\,D$	Series 2 $h = 0.5\,D$	Series 3 $h = 0.63\,D$	$D_1\,\varnothing \times h_1$ in mm
32	6.5 — 12.5	—	16	20	$80 \times 30-35$
40	12.5 — 16	16	20	25	$100 \times 35-40$
50	16 — 20	20	25	32	$125 \times 40-45$
63	20 — 25	25	32	—	$160 \times 45-50$
80	25 — 32	32	—	—	$250 \times 50-60$

speed, when light metal particles will bite to the die edge and cause damage to the surface of the issuing bar or tube. A very great number of
these proposals deal with cooling devices for the die to increase the extrusion rate and the press output. However, such die designs are used in
relatively few cases only, because great difficulties are encountered in
changing such dies, which are moreover very expensive.

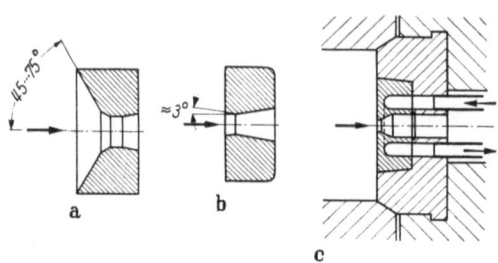

A die, as shown in
Fig. 128, a, is proposed
for the extrusion of hot-
short alloys – such as
AlCuMg – when the tem-
perature of billet, die
and container is to be
300 °C. Ram speeds of
75 to 100 mm/sec are
said to be attainable
with this design.

Fig. 128a-c. Special types of dies

In order to prevent high tensile stresses on the section in the concentric walls of the die at high extrusion rates, a lead-in tapered to about
3° is to be provided, as shown in Fig. 128, b.

A considerable increase in the extrusion rate and elimination of surface cracks when extruding hot-short aluminum alloys is ensured by
using a die cooling device, an example of which is illustrated in Fig. 128, c.

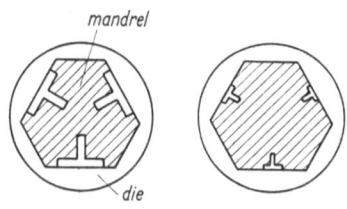

Fig. 129. Die and mandrel for
tapered sections

Shaped sections of gradually or
stepwise changing cross-sectional area
or form are produced using special
means or additional equipment.
Fig. 129 illustrates a mandrel moving
with the ram and being guided in
the die. A number of slots, tapering
lengthwise, are cut in the surface of
the mandrel so that the extruded
section is gradually tapered.

Hollow sections of uniform outer shape, but of continuously or stepwise changing inside shape and reducing cross-sectional area, are made in
a similar simple way, when the mandrel may either move continuously
or may be advanced and arrested, as required, by a mandrel shifting
device.

Stepwise increasing the section is rendered possible by using a number
of dies mounted either in tandem arrangement or inserted concentrically one into the other. Fig. 130 illustrates this method in which after a
given extrusion time – when the shoulder of the mandrel has come close
to the die – the support of the first inner die is removed so that the billet

may be extruded over the thicker part of the mandrel through the second wider die aperture. The first die is left on the section from where it is stripped off later on. Hollow masts are made in this way using a stepped mandrel. The die locking device consists of two opposed slides which traverse the die-carrier. Their operating plungers are laterally mounted on the counterplaten. In their extreme out position the slides release

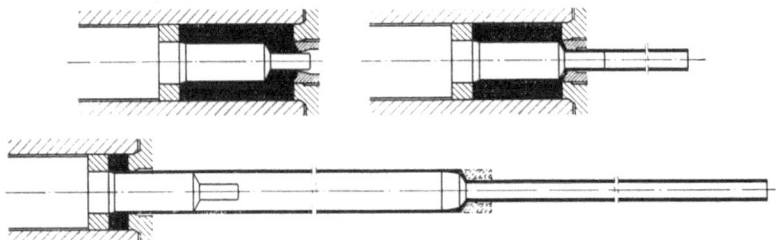

Fig. 130. Tools for extrusion of stepped tubes and sections

the locking of the locking frame so that the die-carrier can only be moved when the die-slides are retracted.

Fig. 131 shows some stepped extrusions which are used as spar junction ends in air frame construction.

Tube and rod extrusion may also be combined with a forging operation. This is clearly illustrated by the examples given in Fig. 132 and 133.

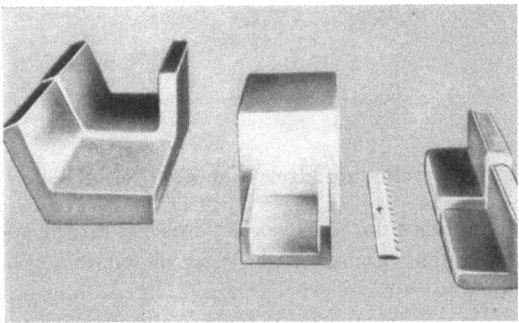

Fig. 131. Stepped sections

In the first stage (see Fig. 132, a) the billet is forced out of the container into a two-part die-chamber the lower of which is closed. This chamber being opened, the shaft is then extruded in a second stage, as seen at b. Fig. 133 shows the forging stage of a hollow body with shoulders (sketch c) in a two-part container and the subsequent extrusion of the tubular portion through a die (sketch b).

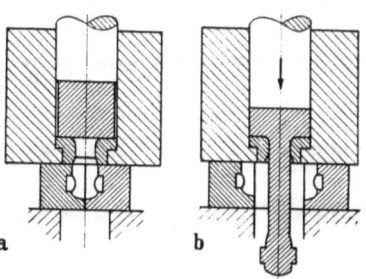

Fig. 132 a and b. Combined forging and extrusion (KREIDLER)

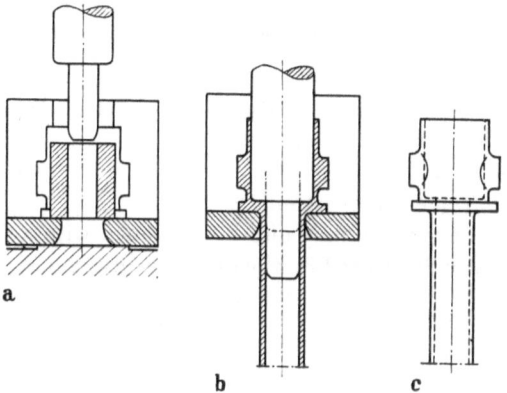

Fig. 133 a–c. Combined hollow extrusion and forging (KREIDLER)

g) Piercing Mandrels and Mandrel-Holders

Piercing mandrels take the heaviest wear of all the tools and therefore give rise to highest expenditure. Wear phenomena occur in the form of contractions and surface cracks caused by cooling stresses and the like, as well as smooth fractures due to over-stressing or incorrect thermal treatment. In case of a smooth fracture which – as a rule – is noticed by the operator by a faint click which should cause him to stop the extrusion cycle immediately, the mandrel remains in most cases stuck in the tube. If, however, such fracture occurs right at the beginning of the extrusion cycle, it may have fatal consequences; it happened, for example, that a torn-off mandrel end of 30 mm diameter and about 200 mm length shot like a projectile through an extruded tube of about 1 m length and seriously injured an operator who was standing at a distance of about 15 m from the press. Accidents of this kind which may be caused by a variety of circumstances, e. g. by stresses due to incorrect thermal treatment, are of course extraordinarily rare and they are only mentioned here to point out that greatest caution is advisable. A number of protective measures were taken to prevent accidents, none of which has, however, given perfect results, because they only impaired the access of the run-out table and obstructed the operators.

The main factor affecting the life of the mandrels, is the quality of the tool steels; high-speed steels were developed which lead to the widespread use of the tube extrusion presses. Valuable data on suitable steels are compiled in Table 13 from data given by ASSMANN[1]). The most important property of the mandrel material is a high strength at elevated extrusion temperatures which are as high as 950 °C for heavy metals and 550 °C for light metals. In order to keep the thermal stresses sustained by the mandrel in tolerable limits, extrusion has to be performed at high speed rates so that contact between mandrel and die is kept to a minimum. The length of the billets used for tube extrusion is therefore considerably shorter than that of billets for rod extrusion. The influence of the billet length on the mandrel life is also shown in Table 13. The short billet length furthermore influences favorably the solid seating of the mandrel, due to which close tolerances in the wall thickness of the extruded tubes are obtained. Heavy metals are extruded at high temperatures at ram speeds of about 100 mm/sec, whereas with most of the light alloys – in order to ensure a smooth surface finish – the rate of extrusion is very limited so that the advantage of the lower extrusion temperature is neutralized by the longer extrusion time.

The mandrel is cooled after every extrusion. This is mostly done by dipping the mandrel into a cooling pipe in which it is sprayed with air, oil or hot water. Thick mandrels – on account of the bad thermal conductivity of the alloy steel – tend to crack under cooling stresses; these may be avoided to some extent by hollow-boring the mandrels. Mandrels of over 30 mm diameter are internally cooled, as illustrated in Fig. 134, when extruding at slow rates.

For this purpose mandrel a and mandrel-holder b are provided with a central bore which accomodates feed pipes c and d. The cooling water is admitted to the hollow space in the mandrel through a number of holes drilled into pipe c and flows back through the mandrel-holder into chamber e of the piercer bar f. The water passage from the ram-holder g to the movable piercer bar f and vice versa takes place through two immersion tubes h arranged side by side, of which only one is shown in Fig. 134.

---

[1]) ASSMANN, H.: Werkstoffeinsatz und optimale Leistungsausnutzung von Preß-dornen für Metallrohrpressen. Metall 1948, H. 7/10.

Table 13. Steels for

Type of Mandrel	Material Extruded	Extrusion Temp. °C	Diam. of Ram mm	Cooling Method	Analysis of Mandrel Steel					
					C	Si	Mn	Cr	W	V
Tube Mandrels	Nonferrous Metals C Cu E Cu Ms 63 Ms 90[1]) So Ms[2])	700 to 950	under 30	oil or air	0.50 0.60	0.20 0.40	0.20 0.40	3.50 4.00	9.00 10.00	0.80 1.00
	Ms 58 Ms 63 Al Bz 4 Al Bz 9 Al M-Bz Ms 63 F 36	700 to 950	over 30	oil or air	0.30 0.35	0.15 0.35	0.20 0.40	2.20 2.50	4.00 4.50	0.50 0.70
	70/29/1 76/22/2 Cu Ni 80/20 70/30 NS 65 12	700 to 950	over 30	water	0.25 0.30	0.20 0.40	0.30 0.50	0.90 1.20	3.50 4.00	0.15 0.25
Tube Mandrels	Light Metals Al 98-99-99.7 Al Cu Mg Al Cu Ni Al Mg Si	320 to 520	under 40	emulsion and air	0.30 0.35	0.15 0.35	0.20 0.40	2.20 2.50	4.00 4.50	0.50 0.70
	Al Mg 3-9 Al Mg Mn Al Mn Mg Al 3-9 Mg Zn Mg Mn	320 to 380	over 40	emulsion and air	0.40 0.50	0.80 1.00	0.20 0.40	1.50 1.80	1.70 2.00	0.15 0.25
Bridge Dies	Mg Al 3-9 Mg Zn Mg Mn	320 to 380	any diam.	none	0.30 0.35	0.15 0.35	0.20 0.40	2.20 2.50	4.00 4.50	0.50 0.70
Tube Mandrels	Zinc Alloys Zn 99-99.97 Zn Al 4 Cu 1 Zn Cu 1	200 to 280	any diam.	emulsion	0.57 0.65	0.70 1.00	0.60 0.90	1.05 1.30	— 	0.08 0.15

[1]) Ms = brass   [2]) So Ms = special brass

and Life of Mandrels

U.T.S. kg/mm²	Temperatures °C				Average Number of Extrusions			Remarks
	Pre-heat.	Quench.	Temper.	Anneal.	Billet Material	Billet length under 250 mm	over 250 mm	
155 to 165	700 to 800	1,120 to 1,140 oil	630 to 650	820 to 840	Cu So Ms	350 to 450	—	Predrilled billet
					Ms 58-63	475 to 625	—	Billet pierced
150 to 160	600 to 700	1,040 to 1,050 oil	600 to 620	760 to 780	Cu So Ms	520 to 700	—	Billet pierced
					Ms 58-63	625 to 880	—	
145 to 160	600 to 700	1,030 to 1,050 water/oil	500 to 540	760 to 780	Cu So Ms	600 to 700	250 to 350	Billet pierced
					Ms 58-63	650 to 800	—	
170 to 180	600 to 700	1,040 to 1,050 oil	570 to 600	760 to 780	Al-Alloys	800 to 1,500	700 to 1,200	Predrilled billet
					Mg-Alloys	600 to 1,100	700 to 1,000	
165 to 177	600 to 700	980 to 1,000 water/oil	520 to 570	760 to 780	Al-Alloys	2,000 to 3,000	1,200 to 2,000	Predrilled billet
					Mg-Alloys	900 to 2,500	600 to 1,500	
150 to 160	600 to 700	1,030 to 1,040 oil or hot water 500/550 GS 430	600 to 610	760 to 780	Mg-Alloys	800 to 1,500	—	Brittle cleavage fractures and deflection of bridge
150 to 160	300 to 400	840 to 860 water/oil	380 to 400	700 to 720	Zn-Alloys	—	—	—

Connecting between the cooling-water pump and the movable ram-holder g is established by two hoses i and j.

The wear resistance of the mandrels depends in the first place on the hardness and the smooth finish of the surface. They are therefore super-finished, i. e. their surfaces are lapped and polished after grinding.

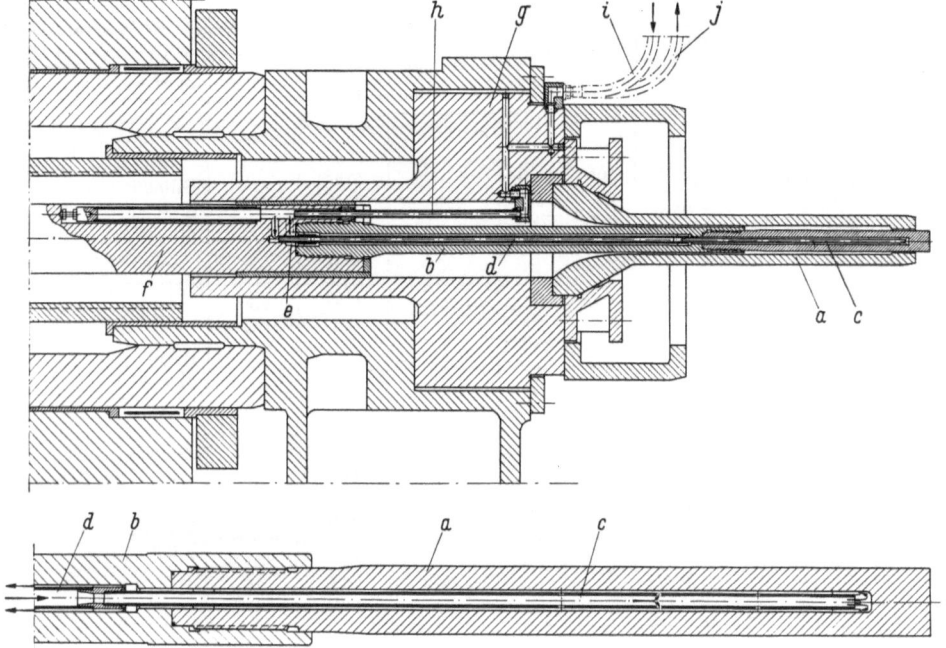

Fig. 134. Device for internal cooling of mandrels

Many of the light alloys have a very high affinity for steel. If particles of the metal of the billet stick to the mandrel, they will heavily attack its surface and soon destroy it. This tendency to "stick" is greatly reduced by employing mandrels of high-grade chromium steel. Another method is to coat them with a protective film made up of a mixture of oil and flaky graphite. At a temperature of 300 to 400 °C the mandrels are dipped in this solution, which then burns into the surface giving it a jet black, glossy color.

In order to minimize the wear of the mandrels, they are well lubricated like all of the other tools which come into contact with the billets, i. e. die, container bore and dummy-block; the lubricant being in most cases a paste made of superheated steam cylinder oil and flaky graphite.

When extruding light metals, lubrication is in many cases not employed so as to maintain a clean surface of the extrusions. On the other hand, however, such lubricants as beeswax, molycote or other commercial grades are used.

The smooth fractures are as a rule caused by mechanical overstressing, if, for example, the billet is not properly upset in the container or if the billet is unevenly heated, in which case considerable bending forces bear upon the mandrel (see Fig. 120). Under such circumstances it may happen that a mandrel withstands 100 to 200 extrusions only, while its life – under more favorable conditions – might be ten times as long. This shows the immense importance of a uniform heating of the billet and of

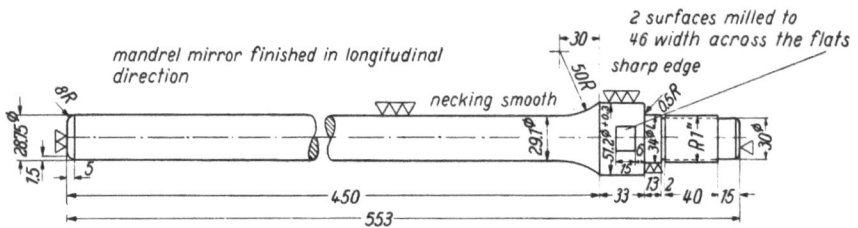

Fig. 135. Mandrel for extrusion of tubes

the skill of the press operator who – while controlling the various motions of the press – must be able to "sense" to what degree the tools may be stressed.

When designing mandrels (Fig. 135) it must be considered that the neckings must be of conical or widely radiused shape so as to prevent the occurrence of stress peaks due to notch effects. The cylindrical part of the mandrel is slightly tapered – about 0.2 to 0.3 mm over a length of 500 mm – so that it will easily detach from the extruded tube; when making this taper the tolerance allowed in the wall thickness has to be considered.

A straight or tapered fine thread is provided in the mandrel-holder to secure the mandrel. The tapered thread which permits of changing the mandrel rapidly, has not found a wide application as the mandrel tends to loosen easily.

The cross-section of the core of the thread has to take heavy tensile stresses both during extrusion and when pulling the mandrel out of the extruded tube. That is why bayonet unions between mandrel and holder are not used, because in such case the diameter of the head of the mandrel-holder would be too wide and the extrusion ram would be weakened too much.

When extruding hollow sections, the mandrel is in most cases fixed rigidly in the extrusion ram (see Fig. 136). The wall thickness of the section is precisely adjusted by checking the cross-section of the annulus

11*

with the help of a template or calipers. Mandrel and die must be adjustable to allow for corrections to be made, which, however, requires special devices.

A die adjusting device in a 600-ton tube extrusion press has already been described in detail on p. 126. In most cases, however, resort is only taken to setscrews arranged directly in the die or ram-holder. The set-

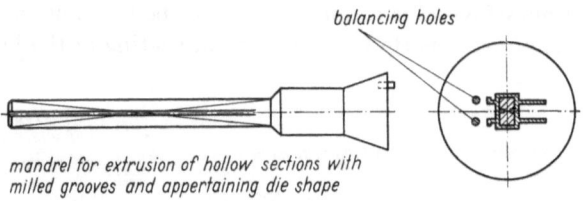

mandrel for extrusion of hollow sections with milled grooves and appertaining die shape

Fig. 136. Mandrel for extrusion of hollow sections (ASSMANN)

screws in the ram-holder serve to adjust the ram or the mandrel. In this case the mandrel is arrested in the die which permits of easily checking the correct adjustment in horizontal presses.

Two centering devices are consequently employed to adjust the tooling for the extrusion of hollow sections: one is used to eliminate eccentricities and the other to eliminate angular differences between mandrel and die.

h) Containers

The container consists of a thick-walled liner which serves to accomodate the billet to be extruded, on which – as illustrated in Fig. 137 – a cylindrical jacket is shrunk. Containers capable of taking highest stresses are made in three parts, i. e. the liner is surrounded by an intermediate sleeve. Shrinking must be carried out with highest care and precision, as it plays an important part in determining the life of the container, the high prime cost of which influences the economy of extrusion essentially. Proper choice of steels is another important factor; details compiled from valuable experience gathered in industrial practice, are shown in Table 14. The liner in particular must possess a high degree of hot-strength and retention of hardness, a high resistance to wear, and it must be capable of withstanding high temperature gradients, which lead very easily to thermal cracks and scoring. Furthermore a high degree of thermal conductivity is required so as to transfer the heat rapidly and to ensure the retention of hardness.

Notwithstanding most accurate choice of steel and most careful treatment, the liner is subject to high wear which affects its life to a very high degree. This wear is highest a short distance from the die. In general the life is 15,000 to 20,000 extrusions depending on the billet temperature and the extrusion qualities of the metal concerned. Under favorable conditions and with easy alloys, such as brass Ms 58 for example, 50,000 and more extrusions may be performed. When the die end of the liner has worn to the permissible limit, the container is turned around and used

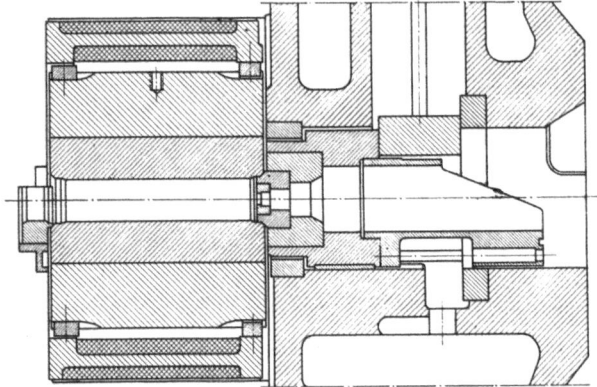

Fig. 137. Two-part container

from the other end or it is bored out to a wider diameter and eventually replaced by a new liner.

The main property required of sleeve and jacket is a high degree of toughness, when on the other hand a lower resistance to wear is acceptable. Steels treated to an ultimate tensile strength of 130 to 150 kg/mm² for liners, 110 to 120 kg/mm² for sleeves and 90 to 110 kg/mm² for jackets are generally chosen. Table 14 shows very clearly the limits for the retention of hardness and for the range of application of various hot working steels with the extrusion temperatures of the most important alloys.

When being extruded, the billets are in a plastic state. The side pressure p_i on the shell surface of the container bore which stresses the container, is smaller than the ram pressure p. In the liquid state the pressure would be equal in all directions and therefore $p_i = p$. Unfortunately there are no precise test results available as yet on the value of the side pressures occurring with different metals, different billet temperatures, ram speeds etc., and calculations are therefore based on $p_i = 0.5$ to $0.8\ p$, which are empirical values; the lower values are applicable to stiff alloys at normal extrusion temperatures.

The question when to prefer two- or three-part containers, cannot be answered off hand. This depends on the stresses occuring and the material chosen. Two-part containers possess a liner with essentially thicker wall and are preferred for the extrusion of heavy metals at high temperatures

Table 14. Limits of the retention of hardness and of the range of application
(After K. LAUE)

Material No.	Mo	V	W	Co	Limit of retention of hardness / Limit of range of application
	Hot working steels Percentage of				Limit of retention of hardness Limit of range of application
2584	0.9	0.7	9.0	–	
2662	–	0.3	9.0	2.0	
2581	–	0.4	9.0	–	
2365	2.8	0.5	–	0.3	
2567	–	0.6	4.5	–	
2343	1.5	0.4	–	–	
2564	–	0.2	4.0	–	
2560	0.4	0.3	2.8	–	
2603	0.5	0.8	0.5	–	
2542	–	0.2	2.0	–	
2242	–	0.1	–	–	

Limits of Extrusion Temperatures

100° 200° 300° 400° 500° 600° 700° 800° 900° 1000°
Lead-a Tin Alloys | Zinc Alloys | Alu-minum Alloys | Brass Alloys | Stiff Cu Alloys
Mg Alloys | Cu-Ni-Alloys

and at high extrusion rates, since in this case the hardness and strength of the material are maintained better, because peak temperatures are compensated more readily by a thick-walled liner. Replacement of the liner after final wear is more cheaply made with three-part containers, since the wall of the liner is thinner. Due to its higher elasticity the shrink assembly is facilitated.

The wall-thickness of the various casings in compound containers are – to begin with – determined from empirical values. According to a rough-and-ready rule the outside diameter of the container is to equal about 4 to 5 inside diameters, taking into account the strength of the material and the ram pressures occurring.

When determining the wall thickness of the container casings, one has to start from the hypothesis that in hollow cylinders the resistance to internal pressures is highest when the ratios between the diameter of the individual casings and the diameter of the liner are equal. The liner of a two-part container would therefore have – at a diameter ratio of $u = d_a : d_i = 4$ – an outside diameter of d_1 which results from $u_1 = d_1 : d_i = \sqrt{u} = 2$. Likewise, the diameter ratio of liner and sleeve in a three-part container would be $u_1 = u_2 = \sqrt[3]{u} = 1.58$. In practice, however, some slight deviations are made from these ratios to allow for the weakening of the outer jackets due to holes being drilled in them to accomodate heaters and to choose equal outside diameters for a number of liners with varying inside diameters d_i to facilitate replacement. It is general prac-

Table 15. *Dimensions for Container Liners and Sleeves (Suggested Standards)*[1])

Liner

Extrusion Pressure $p = \dfrac{P}{F}$ kg/mm²	Outside diameter of Liner D_1 mm	Design of Container
100	B 1.6	
80	B 1.6	3-part
63	B 1.5	
50	B 2	
40	B 1.8	
31.5	B 1.6	2-part
25	B 1.4	

Sleeve

Extrusion Pressure $p = \dfrac{P}{F}$ kg/mm²	Outside diameter of Sleeve D_2 mm	Design of Container
100	D_1 1.8	
80	D_1 1.6	3-part
63	D_1 1.5	

Example for the dimensioning of liner and sleeve on a 2,000-ton tube and rod extrusion press

Press Power P in tons	Container Bore B Diam. mm	Extrusion Pressure $p = \dfrac{P}{F}$ kg/mm²	Liner D_1 Diam. mm	s_1 mm	Sleeve D_2 Diam. mm	s_2 mm	Design	Jacket D_3 Diam. mm
	160	100	265	52.5	450	92.5	3-part	
	180	80	280	50	450	85		as
2,000	200	63	300	50	450	75		calcu-
	224	50	450	113	—	—		lated
	250	40	450	100	—	—	2-part	
	280	31.5	450	85	—	—		

tice to start from a mean inside diameter and to increase the wall thickness of the outer jacket by the dimension of the heater bores. Examples of container dimensions are shown in Table 15.

The diameters of the various casings having been determined on the above principles, the strength of the material has to be calculated.

A closed, thick-walled hollow cylinder being loaded in internal pressure, is submitted to a three-dimensional state of pressure by forces act-

---

[1]) ARENZ, M.: Die Normung der Strangpreßwerkzeuge. Z. f. Metallkde. 46 H. 3

ing in tangential, radial and axial direction; tangential stresses are highest and most dangerous of all stresses occurring. Their maximum value lies in the inner grain and decreases heavily towards the outside. In order to reduce the tangential pull stresses in the innergrain and simultaneously to increase the properties of the outer grain of the material, the container is shrink assembled, whereby the liner is provided with a residual compressive stress and the outer jacket with an initial pull stress.

The radial compressive stresses occurring are highest in the inner fiber, too, and fall off on their way to the outer fiber to zero. They are perpendicular to the tangential stresses and are often disregarded in rough calculations.

Axial stresses occur in closed cylinders only. The container being, however, open on either end and axial frictional forces on the periphery of the container bore as well as other axial forces being low in most cases, it is general practice to consider axial stresses not to be prevailing.

Checking of the wall thicknesses of a container is either done on the theory of elasticity or on the theory of deformation, in which the compound stress – made up of tangential, radial and axial stresses – is referred to a reduced or comparative stress. At the present time the strength of the material is mostly calculated with the help of the theory of deformation, because reference stresses are obtained by this method, which are in good agreement with the tests and slightly higher than the reduced stresses obtained by the theory of elasticity.

Starting formulars after C. BACH, by means of which tangential, radial and axial stresses at any point of an open or closed cylinder under internal or external excess pressure are calculated, are compiled in Table of Formulas 1. These formulas apply to purely elastic deformation only, strictly speaking to deformation up to the elastic limit, approximately, however, up to the yield point.

According to the theory of elasticity the material stress

$$\sigma_{\text{red}} = \sigma_1 - \frac{\sigma_2 + \sigma_3}{m}$$

attains its maximum value where the elongation

$$\frac{\Delta l}{l} = \varepsilon = \frac{1}{E}\left(\sigma_1 - \frac{\sigma_2 + \sigma_3}{m}\right)$$

shows its highest value. This is the case in the direction of the highest principal stress occurring, so that this has to be inserted in the formula as σ_1. As with a shrink-assembled container the highest stress of the individual casings occurs in their central bores, the tangential stresses σ_t, the radial stresses σ_r and, if necessary, also the axial stresses σ_l which

Table of Formulas 1. *Stresses in thick-walled cylinders*

$u = \dfrac{d_a}{d_i}$ d_a = outside diameter
d_z = diameter at any point
$u_x = \dfrac{d_a}{d_x}$ d_i = inside diameter

Compressive stresses, also internal pressure p_i and external pressure p_a, have the negative sign, pulling stresses the positive sign.

	At internal pressure p_i ($p_a = 0$)	At external pressure p_a ($p_i = 0$)
For any point in the cylinder wall	$\sigma_{t_x} = -p_i \dfrac{u_x{}^2 + 1}{u^2 - 1}$	$\sigma_{t_x} = +p_a \dfrac{u^2 + u_x{}^2}{u^2 - 1}$
	$\sigma_{r_x} = +p_i \dfrac{u_x{}^2 - 1}{u^2 - 1}$	$\sigma_{r_x} = +p_a \dfrac{u^2 - u_x{}^2}{u^2 - 1}$
For the inner grain	$\sigma_{t_i} = -p_i \dfrac{u^2 + 1}{u^2 - 1}$	$\sigma_{t_i} = +p_a \dfrac{2\, u^2}{u^2 - 1}$
	$\sigma_{r_i} = +p_i$	$\sigma_{r_i} = 0$
For the outer grain	$\sigma_{t_a} = -p_i \dfrac{2}{u^2 - 1}$	$\sigma_{t_a} = +p_a \dfrac{u^2 + 1}{u^2 - 1}$
	$\sigma_{r_a} = 0$	$\sigma_{r_a} = +p_a$
σ_l for every point in the cylinder wall	$\sigma_l = -p_i \dfrac{1}{u^2 - 1}$	$\sigma_l = +p_a \dfrac{u^2}{u^2 - 1}$

σ_l occurs in closed cylinders only.

occur at those points, are calculated with the help of the initial formulas tabulated in Table 1, all of which being separately calculated for the individual and simultaneously acting loads, such as internal pressure p_i, external pressure p_a (due to the shrink pressure between liner and sleeve and the shrink pressure between sleeve and outer jacket), additional axial stresses etc. The tangential stresses σ_t thus found in one point are added up – according to their signs -- to $\Sigma\sigma_t$, the radial stresses σ_r to $\Sigma\sigma_r$ and the axial stresses σ_l to $\Sigma\sigma_l$. The highest of the determined combined stresses is marked σ_1, the others σ_2 and σ_3. From these values the material stress $\sigma_{1\,red}$ is determined for the point considered according to the above formula. In most cases POISSON's ratio is inserted as $m = 10/3$. $\sigma_{1\,red}$ must not exceed the permissible value $\sigma_{z\,zul}$ or $\sigma_{d\,zul}$[1]).

According to the theory of deformation, the material stress which is derived from the deformation energy, has its maximum value at that point where the comparison stress

$$\sigma_c = \frac{1}{\sqrt{2}} \sqrt{(\sigma_t - \sigma_r)^2 + (\sigma_r - \sigma_l)^2 + (\sigma_l - \sigma_t)^2}$$

[1]) $\sigma_{z\,zul}$ = permissible pull stress; $\sigma_{d\,zul}$ = permissible compressive stress.

is at its maximum. With a shrink-assembled container the combined stresses $\Sigma\sigma_t$, $\Sigma\sigma_r$, $\Sigma\sigma_l$ at the bores of the individual casings are therefore calculated, as indicated above for the theory of elasticity, and inserted in the formula for the comparison stress, when σ_v must again not exceed the permissible value $\sigma_{z\,zul}$ or $\sigma_{d\,zul}$.

When the reduced stresses σ_{1red} or the comparison stresses σ_v at the bores of the individual casings correspond with the permissible stresses of the materials employed, then the most favorable strain of the material has been attained. To attain calculated stresses at the critical points calls for correct determination of the shrink pressure between the individual casings, which in most cases is only attained after repeated correction of the values first assumed. The shrink pressure is produced by heating the outer casing and assembling it over the inner casing whose diameter has been increased by the amount of contraction of the former. After equalization of the temperature the inner casing is therefore compressed and the outer expanded. Consequently, the amount of contraction s equals the algebraic sum of the decrease in diameter Δd_k of the inner casing and the increase in diameter Δd_g of the outer casing at their interface:

$s = \Delta d_g - \Delta d_k$, when Δd_k is negative.

In general $\Delta d = d\,\dfrac{\sigma_{t\,red}}{E_t}$, when $\sigma_{t\,red} = \sigma_t - \dfrac{\sigma_r + \sigma_l}{m}$ is the tangential stress resulting from the shrink pressure p_s and E_t = modulus of elasticity at operating temperature.

With $\sigma_l = 0$ (assumption for open cylinder) and $m = 10/3$ (POISSON's ratio), $\sigma_{t\,red} = \sigma_t - 0.3\,\sigma_r$,

hence $\quad s = \dfrac{d_g}{E_{t_g}}\,(\sigma_{t_g} - 0.3\,\sigma_{r_g}) - \dfrac{d_k}{E_{t_k}}\,(\sigma_{t_k} - 0.3\,\sigma_{r_k})$.

However $\sigma_{r_g} \equiv \sigma_{r_g}$ (shrink pressure p_s),
further $d_d \cong d_k = d$ (common contact diameter)
and with uniform distribution of temperature $E_{t_g} \cong E_{t_k}$ may be inserted, the equation for the amount of contraction s is simplified to:

$$s = \frac{d}{E_t}\,(\sigma_{t_g} - \sigma_{t_k}).$$

Amounts of contraction thus obtained are in the range of $1/300$ to $1/500$ of the diameter. When, however, a temperature gradient is anticipated in the container and hence a change of the radial loads, the calculation of the amount of contraction is no longer possible by the same simple method. In practice, an empirical value of about $1/500$ to $1/600\,d$ is often employed.

Fig. 138 illustrates the tangential stresses in a three-part container both in loaded and unloaded condition. When determining the permissible stresses $\sigma_{z\,zul}$, these are kept – for safety reasons – about 15 to 20%

below the hot elastic limit, provided one has not to disconsider this safety
factor partly or completely because of particularly high internal pressures.
This is permissible because of the uneven distribution of stress over the
cross-sectional areas of the rings[1]). Attaining the hot elastic limit does
not necessarily result in a fracture, but it may lead to local deformations
which would reduce the life of the container.

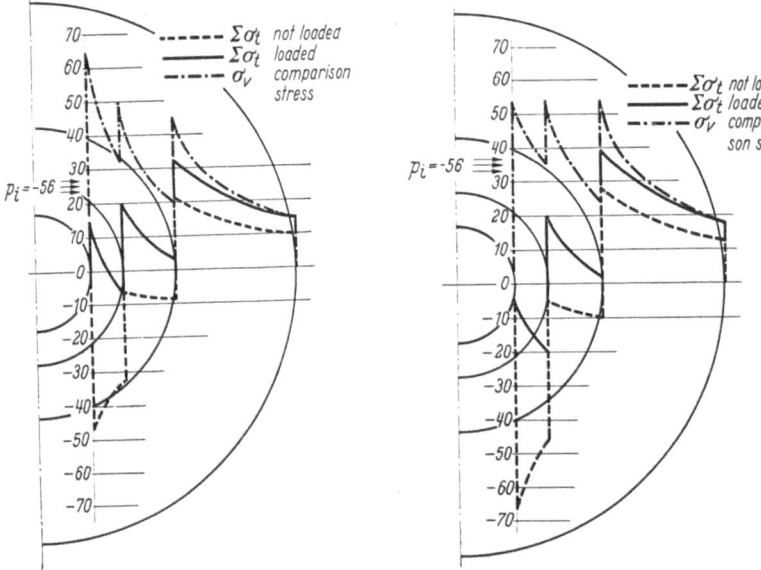

Fig. 138. Stress distribution in three-part
container with slightly prestressed liner

Fig. 139. Stress distribution in three-part
container with heavily prestressed liner

It has further to be considered that liner and sleeve are subject to a
stress reversal and the outer jacket to a repetitive stress. The stress
reversal of the liner may cause small hairline cracks in the inner surface
which may result, for example, from temperature fluctuations, to widen
somewhat during the changing from compression to pull. The extrusion
of alloys containing lead and bismuth is liable to cause further cracking,
as these alloying components penetrate into the grain boundaries of the
steel. This can be prevented by applying a very high pre-tension to the
liner by heavy shrinking which eliminates any tangential pulls in the
liner even when the container is loaded. Such distribution of stress is
illustrated in Fig. 139. In this case the liner is only subject to a repetitive
stress which, however, results in a correspondingly higher stress of the
sleeve and the jacket.

---

[1]) MAIER, A. F.: Die Beherrschung von hohen Drücken bei Gefäßen. Techn.
Mitteilungen Krupp, Okt. 1937, H. 7.

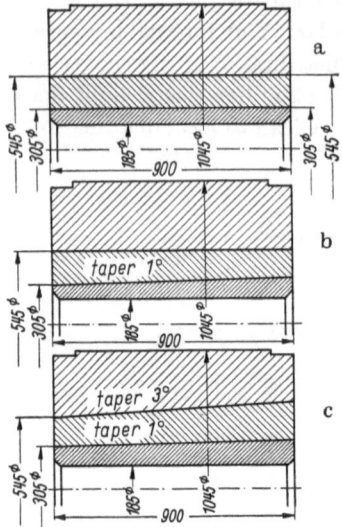

Containers are constructed both with cylindrical and tapered liners and sleeves (see Fig. 140). Cylindrical liners and sleeves are advantageous in so far as replacement is made simpler and truer to dimensions, whereas tapered liners facilitate their exchange. Tapers generally chosen are 1° for the liner and 3° for the sleeve. The tapered liner is in some cases exchanged during operation when changing over to another billet diameter, thus eliminating the need to keep a number of containers in stock. This, however, requires a high degree of accuracy and a lot of practical experience.

In order to keep away the ram pressure from the shrink fits when ejecting a

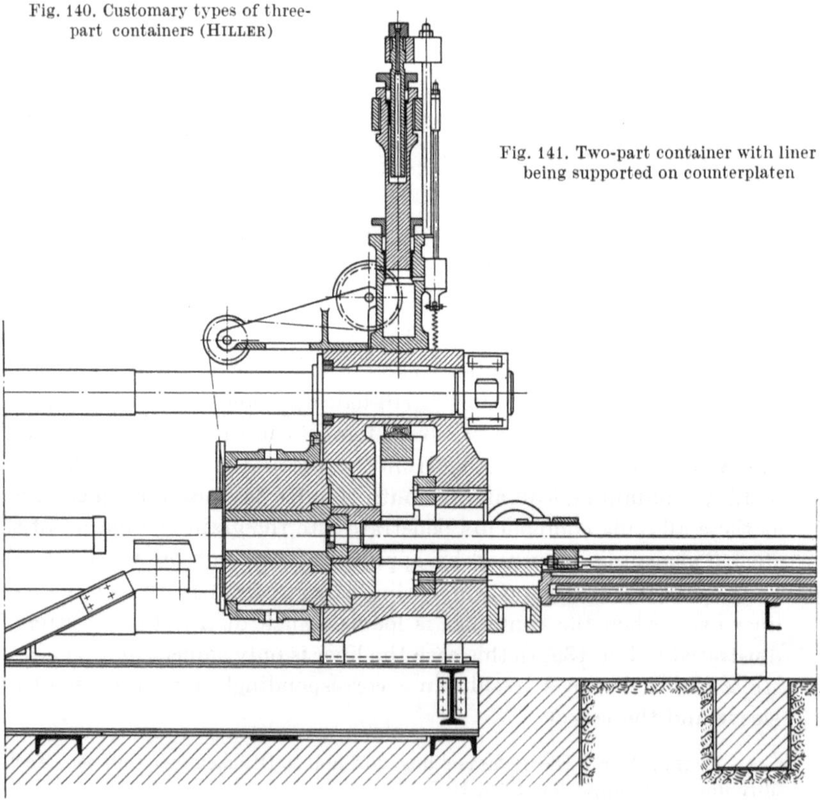

Fig. 140. Customary types of three-part containers (HILLER)

Fig. 141. Two-part container with liner being supported on counterplaten

"sticker", it has proved advantageous to have the liner abut against the counterplaten by means of a collar, as illustrated in Fig. 141.

The container length L is:

$$L = (l_1 + h) + t + s.$$

l_1 is the billet length; it depends on the maximum billet diameter d and is to be $l_1 = 2.5$ to $3.5\ d$. When extruding tubes the value $l_1 = 2.5\ d$ is not to be exceeded because of wear and deflection of the mandrel. h is

Table 16. Dimensions of Dummy-Blocks
(Suggested Standards)[1])

Diameter D Nominal[2]) Size in mm	Thickness of Dummy-Block s in mm			
	Series 1 (0.4 D)	Series 2 (0.45 D)	Series 3 (0.50 D)	Series 4 (0.56 D)
90		40	45	50
95		40	45	50
100		45	50	55
106		45	50	55
112		50	55	65
118		50	55	65
125		55	65	70
132		55	65	75
140		65	70	80
150		65	75	85
160		70	80	90
170		70	85	95
180		80	90	100
190		80	95	105
200		90	100	110
212		95	105	115
224		100	110	125
236		105	115	130
250		110	125	140
265		115	130	150
280		125	140	160
300	120	130	150	170
315	125	140	160	180
335	130	150	170	190
355	140	160	180	200
375	150	170	190	210
400	160	180	200	225
425	170	190	210	235
450	180	200	225	250
475	190	210	235	265
500	200	225	250	275

[1]) See footnote page 167.

[2]) True size depending on container bore and thickness of shell.

Table 17. Dimensions of Container Bores
(Suggested Standards)[1])

Press Power P in tons	Container bores d_i in mm at extrusion pressure $p = P/F$ in kg/mm² (values expressed in round standard figures).												
	100	90	80	71	63	56	50	45	40	35.5	31.5	28	25
630	90	95	100	106	112	118	125	132	140	150	160	170	180
710	95	100	106	112	118	125	132	140	150	160	170	180	190
800	100	106	112	118	125	132	140	150	160	170	180	190	200
900	106	112	118	125	132	140	150	160	170	180	190	200	212
1,000	112	118	125	132	140	150	160	170	180	190	200	212	224
1,120	118	125	132	140	150	160	170	180	190	200	212	224	236
1,250	125	132	140	150	160	170	180	190	200	212	224	236	250
1,400	132	140	150	160	170	180	190	200	212	224	236	250	265
1,600	140	150	160	170	180	190	200	212	224	236	250	265	280
1,800	150	160	170	180	190	200	212	224	236	250	265	280	300
2,000	160	170	180	190	200	212	224	236	250	265	280	300	315
2,240	170	180	190	200	212	224	236	250	265	280	300	315	335
2,500	180	190	200	212	224	236	250	265	280	300	315	335	355
2,800	190	200	212	224	236	250	265	280	300	315	335	355	375
3,550	212	224	236	250	265	280	300	315	335	355	375	400	425
5,000	250	265	280	300	315	335	355	375	400	425	450	475	500

Fig. 142. Container with split holder

[1]) See footnote page 167.

the elongation of length l_1 during piercing of the billet by the maximum mandrel.

Dimension t is the length over which the die protrudes into the container and s the thickness of the dummy-block which is chosen to be $s = 0.4$ to $0.6\ d$. Empirical values of dummy-block thicknesses are tabulated in Table 16; the series to be chosen depends on the thermal stress.

Efforts have been made to standardize container dimensions and suggestions have been made for the standardization of container bores in dependence of the press tonnage, which are shown in Table 17 (see also Fig. 112, p. 132).

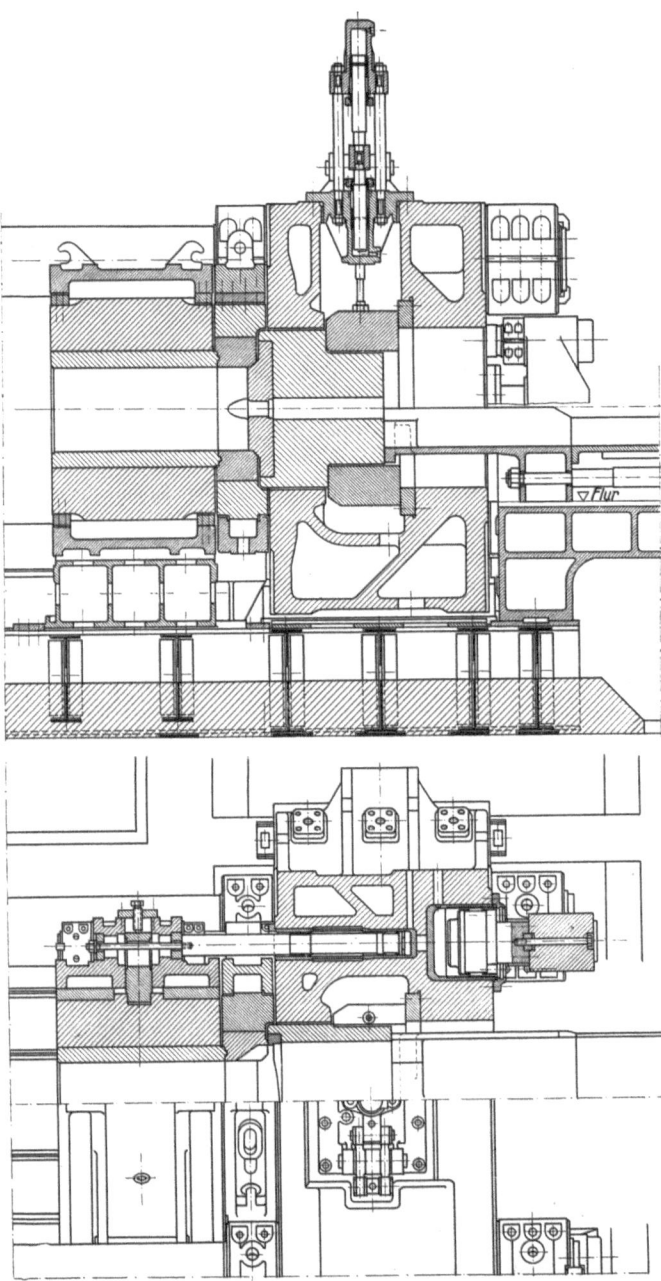

Fig. 143. Container and adapter with lateral flarings for production of slabs

With older rod and tube extrusion presses the container is in many cases set in the center of the counterplaten. This design has, however, proved unsuitable because of the high transmission of heat. The method presently used is a separate container-holder, either bolted to the counterplaten or used to shift the container. The container-holder is made of a steel casting and is insulated both on the inside and outside so as to prevent absorption and reflection of heat. It has furthermore proved to be very advantageous to support the container both in the horizontal and vertical plane on four ribs, as illustrated in Fig. 142, to allow for a free expansion due to heat changes.

The container-holder, as per Fig. 142, is in two-part design. The higher expenditure incurred, which is even increased by the unfavorable transmission of the displacement forces, is made up for by the advantage of the container being changed more easily. With this design the container is lifted out through the top which must not be obstructed by a press column.

Movable container-holders are provided with adjustable guide shoes to allow for exact alignment of the axes of container and press. The container-holder is further designed so as to permit of swiveling the container 360° as the liner wears more quickly on the die end than on the charging end. This of course calls for die centering- and sealing cones to be provided on either end of the liner.

Apart from these standard containers, there are also some special designs. Fig. 143 illustrates, for example, a container provided with a ring on its rear end which is used for the extrusion of slabs, whose width is greater than the diameter of the billets and which are ultimately rolled down to strips on a rolling mill. The ring is surrounded by a band and is set in a cast plate which is secured to the container-holder; it has the same diameter as the liner and is provided with lateral recesses. The contact between the die and these recesses is made by surface, whereas over the rest of the periphery it is established by a sealing taper. Sealing between ring and container is ensured by groove and tongue.

There has also been some publication on special containers, the cylindrical bores of which are lined with inserts so as to allow for the extrusion of rectangular copper billets.

i) Container Heating Systems

The life of a container depends to a very appreciable extent on the type of heating employed. During the first decades rod and tube extrusion presses were mainly used for the working of heavy metals, for which it is necessary to heat the container to a temperature of 300 to 400 °C. Once in operation the container is warmed by the passage through it of successive hot billets. This may raise its temperature to a degree that makes it necessary to cool the container with warm air between the individual extrusion cycles.

The simplest means of heating the container is by introducing hot billets into its bore. This, however, must be done with utmost care as great temperature gradients may occur in the bore through the sickle-shaped air slit, due to which the so-called scoring cracks occur which lead to a rapid destruction of the liner. In order to reduce the heating period, an external heating method was adopted, in which, as shown in Fig. 51, a furnace, burning coke or other solid fuel, was placed under the container-holder and the waste gases were exhausted through a pipe mounted on top of the holder. This was, however, not a perfect solution – in particular with heavy presses – as the liner was heated too slowly and insufficiently.

A substantial improvement was obtained by mounting gas- or oil burners in the container-holder. The heating cycle could thus be sped up and controlled, having on the other hand the disadvantage that the heating of the container was not very uniform and that the hot gases escaping from the slits of the bore of the holder, obstructed the operation of the press.

It was not until the use of electric resistances, mounted in the container-holder in the form of insulated rods, wire or cylindrical coils, that remarkable progress was achieved. The heating system was divided in various opposite panels, thus enabling to control the heating capacity in steps. These systems had, however, the disadvantage that short-circuits occurred frequently. These were caused by the mechanical action of the fine scales detaching from the container jacket and by the fracture of heating conductors due to brittleness and shocks occurring, for example, when the container-holder makes contact in its extreme end position. These troubles diminished in the course of time after the bare heating conductors had been replaced by heating tubes. In these tubes an insulating compound is cast around the conductors thus protecting them against air or oxigen.

With the advent of the extrusion of more difficult light alloys in rod and tube presses, frequent trouble was experienced in that liners and even jackets suddenly cracked. In nearly all of these cases it could be concluded that the capacity of these heating systems – although sufficient for heavy alloys – was absolutely insufficient for light alloys. In any case billets of heavy alloys are hotter than the liner and therefore call for a cooling of the container during continuous operation, whereas the temperature of a light metal billet may occasionally be considerably lower than that of the liner. Most light alloys remain in the container for a considerable time – the extrusion cycle in medium capacity presses occupying as much as half an hour – during which the billet consumes a considerable amount of heat of the liner, whose temperature is higher than that of the billet, which causes part of the shrinkage pretension, i. e. the supporting effect of the jacket, to be eliminated. Finally the liner will crack under the overstressing when the jacket will be fully exposed to the forces of the internal pressure.

In order to render the heating system more efficient, the heat source was no longer placed in the container-holder, but in the jacket proper.

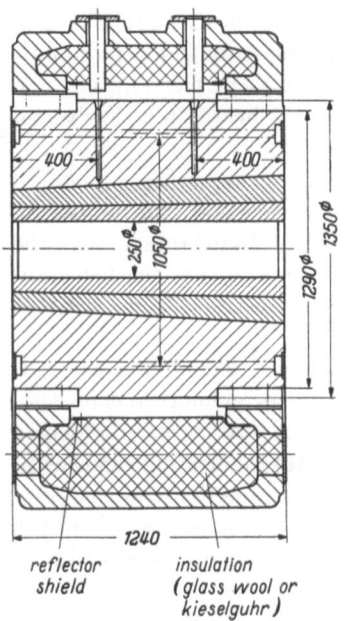

reflector   insulation
shield   (glass wool or
                 kieselguhr)

Fig. 144. Three-part container with bores to accommodate heating and temperature measuring means (LAUE and ARENZ)

For the purpose of preventing heat transfer to the outside, the holder is provided with chambers being filled with an insulating compound and covered with reflector shields (Fig. 144). Placing the heating system even closer to the billet, that is into the sleeve or liner, might seem to be worth aiming at; this would, however, not be expedient with a view to tangential stresses.

The jacket heating is of the resistance- or induction type. In either case longitudinal holes to accommodate the heaters are drilled in the container jacket, as shown in Fig. 144.

In the resistance heating system groups of highly heat-resisting heaters are wired in series and connected parallel to the existing electric circuit. Provision must be made that – should one of the heating coils fail – only a small portion of the complete heating may fail on account of the coils being connected in series and that this failure should be distributed over the full jacket. The heaters are provided with coils of CrNi-steel being embedded in Mg-oxide and placed in V2A-steel tubes. Mainly *Backer* and *Makro* tubes are employed. *Backer* tubes contain a thin heater wire and their temperature at the wall of the tube is in the region of about 800 °C. *Makro* tubes are provided with a number of coils of thicker wire being connected in series. The temperature at the wall of these tubes is somewhat lower and therefore they require a larger heating surface. The temperature is controlled by thermocouples which ensure an intermittent operation of the heating system.

With the induction type of heating insulated copper rods are mounted in holes drilled into the jacket and connected in series, through which alternating current of high intensity is passed. The heat is produced by induced current in the material surrounding the holes and forming a short-circuited secondary circuit. Low-voltage current, controllable, for example, from 20 to 30 volts, is taken from a transformer. The temperature is controlled by thermocouples. The heating rods are more rugged

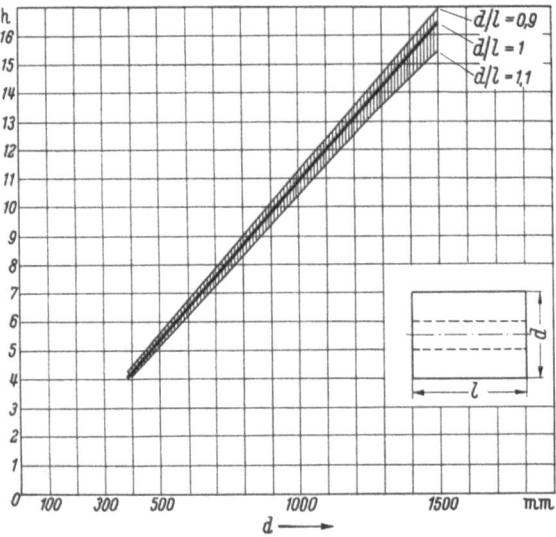

Fig. 145. Heating time in dependence of container diameter (Brown Boveri, Mannheim)

as compared with the resistance heating tubes; however, the efficiency of this heating is lower and additional equipment (transformer and condenser bank) is relatively expensive. Due to the large cross-sections of the conducting wires, much space is required for the supply of power.

Heating cycles and power inputs are tabulated in Fig. 145 and 146. Induction heatings for containers are preferably designed in accordance with standardized inputs: in single-phase connection 10, 16, 25, 40 and 63 kVA, in three-phase connection 100, 125, 160 and 200 kVA.

Advantages and disadvantages of the two systems have to be evaluated in every case, when the initial cost and power consumption play a decisive part.

Heating capacity is determined by practical experience. General requirements are to raise a medium-sized container from room temperature to operating temperature of 400 to 450 °C within about 24 hours.

The containers must be heated at very slow rates and with greatest care because of the low heat conductivity of the alloy steels. A container which is not even loaded, is liable to crack as a consequence of the temperature gradient alone; the temperature difference between two control points should therefore never be more than about 100 °C. The temperature is controlled by a number of thermocouples being arranged at various points of the container; these may also be used for the afore-mentioned switching of the heating current which is cut automatically as soon as the temperature difference in the container exceeds a given value.

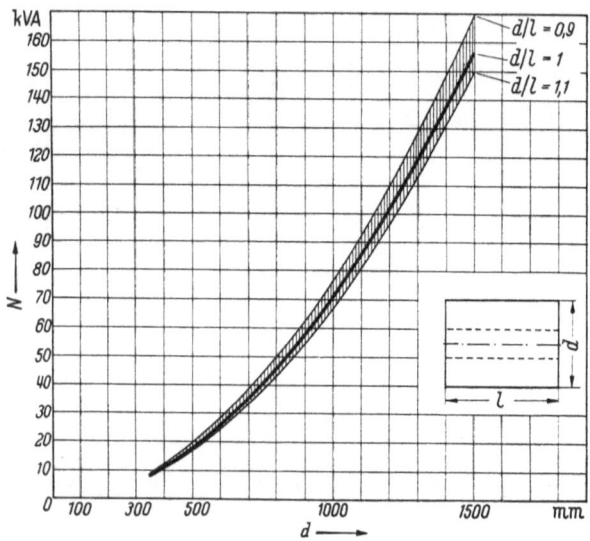

Fig. 146. Connected load in dependence of container diameter (Brown Boveri, Mannheim)

In order to reduce the heating period in the event of the container having to be exchanged, it is appropriate, in particular with heavy press installations, to have the container warmed up in a muffle furnace. If the containers have to be changed often, they are not allowed to to cool down at all, but are kept in the oven at a temperature of about 300 °C.

The largest containers constructed up to date show an outside diameter of about 2 m, a length of about 2.2 m and a bore of 400 to 800 mm. As temperature gradients may in this case occur in longitudinal sense, too, heating cartridges are introduced into the container from both ends and divided up into several heating zones.

In order to minimize radiation losses in the container jackets, it is recommended to also install a resistance heating in the container-holder of heavy presses.

j) Extrusion Rams

The billet is extruded by a solid extrusion ram of circular cross-section or a hollow ram of annular cross-section, as illustrated in Fig. 147, which has to sustain compression and buckling stresses. The yield point of the best steels is now about 120 kg/mm², from which it may be concluded that – when extruding rods by means of a solid ram and extruding tubes by means of a hollow ram – extrusion pressures of 12,000 and

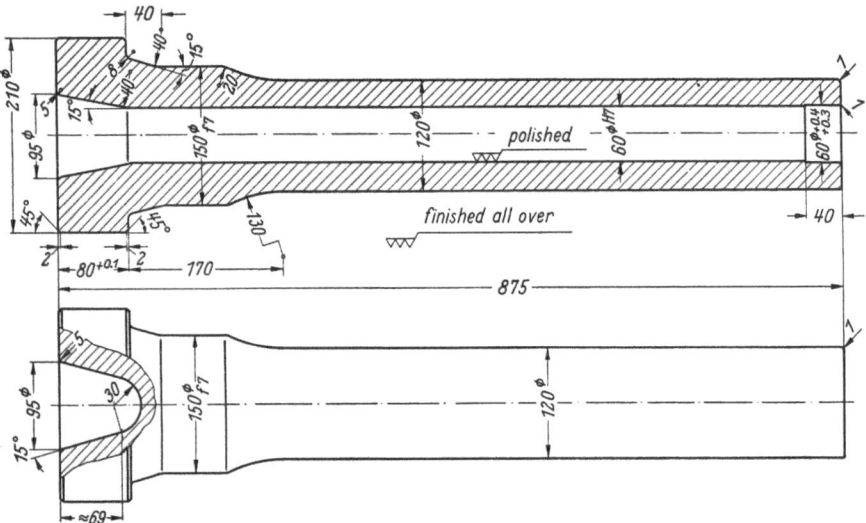

Fig. 147. Tubular ram and solid ram for tube and rod extrusion press

10,000 kg/cm² on the full cross-sectional area of the billet must not be exceeded. Thus the largest possible dimensions of the central bore of the hollow ram in which the mandrel-holder moves, as well as the dimensions of the thickest mandrel and the largest inside diameter of the tube to be produced, are governed by these factors.

A dummy-block is placed against the face of the extrusion ram. For the extrusion of rods the dummy-block is centered on a bolt being inserted in the solid ram and for the extrusion of tubes by means of a hollow ram, on the mandrel. The transition from the cylindrical part of the ram to its flange takes a tapered form. The bearing surface of the flange is dimensioned for a permissible specific unit pressure of about 4,000 kg/cm². The area of the thrust ring which transmits the ram power to the cross-head, is dimensioned for a surface loading of about 1,500 to 2,000 kg/cm². On account of its stress-concentration index the surface and the central bore of the ram are polished and changes in its cross-section are of slim and heavily radiused shape. Slight turning marks are particularly liable to cause fatigue fractures.

When extruding billets "without shell", extrusion rams to which the dummy-block is rigidly attached, are often used, in which case there is only little clearance provided in the container bore. The dummy-block is suitably fixed by a heavy truss bolt or threaded bolt, as shown in Fig. 148, which has to be dimensioned so as to be able to sustain the full pullback power which will stress it up to its yield point. As a rule connection between ram and ram-holder or crosshead is established by means of a flange.

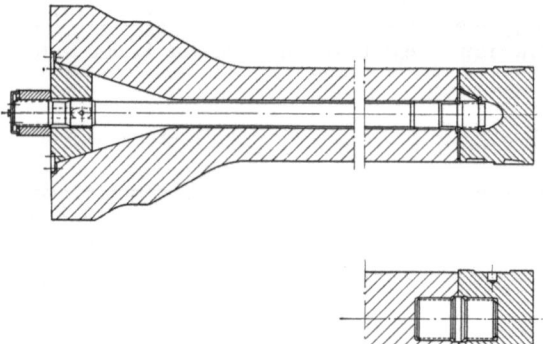

Fig. 148. Ram with screwed-on dummy-block for extrusion of billets without shell

In tube presses in which the die is fixed, the ram-holder is suitably provided with a centering device consisting of setscrews or wedges, which serves to precisely set the mandrel in the die prior to extrusion. As mandrel-holder and mandrel-bar join in this motion, their connection to the piercer bar must be provided with a corresponding clearance transverse to the axis. In addition to this centering device provision is often made to allow for rotating the ram with the mandrel so as to be able to set the proper angular position when extruding hollow sections with the help of an angular mandrel which is rigidly screwed in the ram. Simple setscrews are fully sufficient for this purpose (see p. 164).

k) Die-Carriers, Lateral Die Slides and Rotating Die-Holders

The die-carrier serves to accomodate the die and the die-holder and transfers the full extrusion power to the counterplaten via a locking wedge or -sliding bar (see Fig. 137). It takes the form of a cylindrical tool carrier and is provided with a tubular extension, on which the plunger of the shifting device engages. This is the oldest and best known design. With this die-carrier, in particular in heavy presses, the disassembly of the dies in axial direction, for which an ejector rod was necessary, was rather complicated. Its design was soon improved, taking now the form of a tray, as shown in Fig. 119, in which the die-holder with die as well as the backer rings are placed from above either manually or with the help of a crane.

Preference has recently been given in small- and medium-sized presses up to about 2,500 tons capacity to a lateral die slide, illustrated in Fig. 149, on account of which the delivery end of the press is cleared thus allowing for the installation of cooling-, coiling-, sawing- or other equip-

Fig. 149. Combined 2,500-ton tube and rod extrusion press in three-column design with lateral die slide (By Hydraulik, Duisburg)

ment. Moreover it helps to cut "idle time", as may be seen from the comparisons made in Table 18. This design (see p. 109) has for a long time been disregarded and the main impetus leading to the use of this lateral die slide arose from the urgent requirement for fully or semi-automatic operation and increase in the efficiency of the press.

With heavier presses these advantages lose some of their significance and even some disadvantages become apparent. The main drawback being that the size of the lateral slide and the distance over which it would have to be displaced, would be out of any proportion. When, for example, channel- or I-sections with wide flanges are extruded, the cantilever tongues of the die must be well supported to prevent their tearing due to the bending stresses occurring. The required backer rings are very easily accommodated in a die-carrier, while it would be impossible in most cases to mount parts of such height in a lateral die slide. Furthermore a die slide is unsuitable when two dies are used, for example, for the extrusion of stepped pipes, as illustrated in Fig. 130.

Table 18. Comparison of cycle times of two 1,000-ton rod extrusion presses of different design

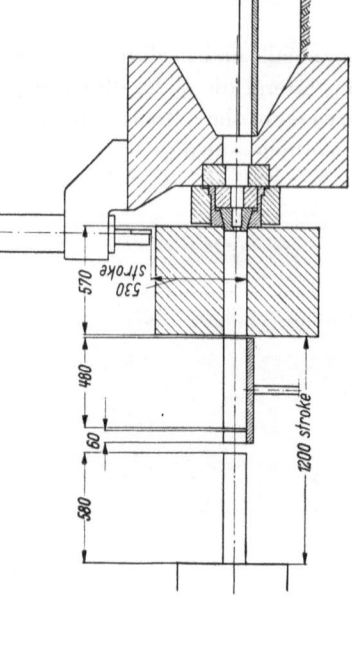

a) With die-carrier and locking sliding bar

	Operation	Stroke mm	Time seconds
1	Upsetting of billet	30	1.5
2	Extrusion	440	10.0
3	Return of ram and container	20	2.0
4	Locking sliding bar out	360	2.5
5	Ejection of discard	150	3.5
6	Die-carrier out	920	4.0
7	Shearing	320	3.0
8	Return of shear	320	2.5
9	Die-carrier in	1,050	4.5
10	Locking sliding bar in	360	2.5
11	Return of ram (during operations 6 to 10)	1,180	(4.6)
12	Loading arm in (during operations 6 to 10)	700	(3.5)
13	Advance of ram (during operations 6 to 10)	600	(3.5)
14	Container closing	20	1.5
15	Loading arm out	700	3.5
	Cycle Time		41.0

b) With lateral die-slide

	Operation	Stroke mm	Time seconds
1	Upsetting of billet	30	1.5
2	Extrusion	440	10.0
3	Stripping of container and ejection of discard	130	3.0
4	Return of ram and container	70	2.5
5	Shearing	530	4.0
6	Return of shear	530	3.0
7	Return of ram (during operations 5 and 6)	1,070	(4.0)
8	Closing of container	200	2.5
9	Loading arm in	700	3.5
10	Advance of ram	600	3.5
11	Loading arm out	700	3.5
	Cycle Time		37.0

The lateral die slide – as a rule – is provided with two openings, one serving to accomodate the die and the die-holder and the other permitting of ejecting the shell and the ejector disk. "Stickers" may also be ejected through this opening. The die slide having been moved sideways and the die being in the outer station, the die may easily be cleaned and, if necessary, cooled and lubricated.

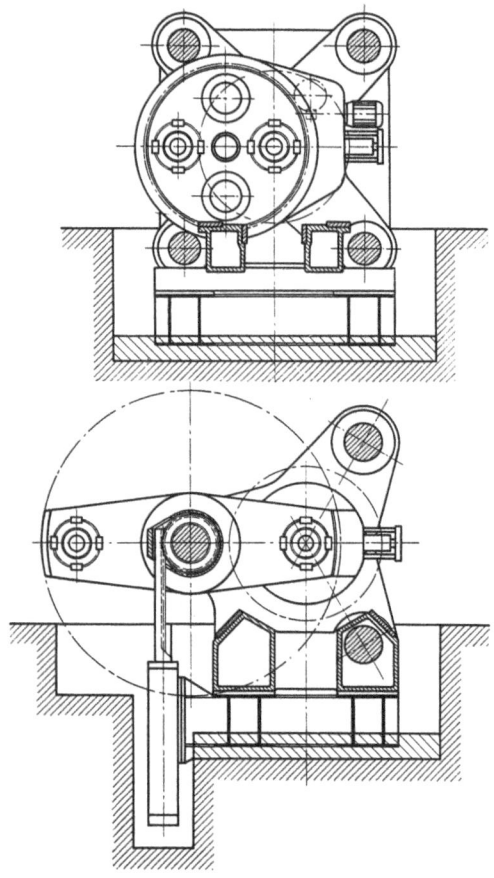

In the outer station the die is replaced by simply lifting it out of the open station. If this is of the enclosed type, the die is pushed out horizontally by a small piston, whose cylinder is attached to the counterplaten. The slideway for the lateral die slide is often provided with adjustable gibs which permit – in connection with an adjustable stroke limitation – of centering the die vertically and horizontally, thus rendering unnecessary a special ram centering device.

Separation of the discard by means of a saw or shear is done in front of the counterplaten. If the slide is provided with a closed station to accommodate the die-holder, the shear is in

Fig. 150. Rotating disk and rotating arm for four- and three-column press respectively

most cases mounted in the base frame; the shearing blade thus moving upwards. If the station is, however, open at the top, the shear is arranged on the counterplaten and the blade cuts downwards.

When extruding billets at elevated temperatures over long periods – for example when making copper rods and wire – it is advantageous to change the die after each extrusion. In order to accelerate this operation, a revolving disk or arm with two or four openings may be provided behind the die slide in its outer position, which permits of pushing the used

die with its holder into one of the empty openings and after 90 or 180°
rotation, of pushing another prepared and lubricated die rapidly into the
die slide. The used die may then be cooled and redressed without having
to stop the extrusion cycle.

Die slides with an intermediate station for "sticker" ejection
and two die stations which are alternately traveled out to the left- or
right-hand side of the press, call for operation on either side of the
press.

In recent years presses have been built which are provided with a revolving disk or revolving arm on the counterplaten, as shown in Fig. 150, instead of a lateral die slide. The revolving disk is provided with two die-
and ejector stations each, being arranged diametrically opposite each
other, whereas the revolving arm is provided with two die stations only.
The revolving arm is suitable for small and medium presses in three-
column construction, in which the central horizontal press column is
employed as the pivot for the revolving arm. In heavy presses preference
is given to revolving disks which – in the four-column construction –
may be constructed with relatively small outside diameters. The die-
holder must be seated so as to permit a shifting of the axis of rotation
due to the thermal expansion of the counterplaten.

1) Counterplatens

The compression forces emanating from the ram are transmitted to
the counterplaten which is in most cases made of a steel casting and is
subject to bending stresses. Efforts are made to keep its overall length
to a minimum and to provide it on its rear side with a wide delivery
opening so as to permit of easily checking the die position when extruding
tubes or hollow sections and locating flaws in the surface finish right at
the beginning of the extrusion. Ease of access is further desirable when
trying out new dies for complex shaped sections. It often occurs that the
flow of the bar is bad on one side thus causing the bar to be bent. In such
case the bar may be gripped and bent aside with a pair of tongs, whereby
the die will run in. Removing and redressing the die would occupy too
much time. In order to reduce the length of the platen, the sliding bar
or wedge for locking the die-carrier is arranged in the neutral axis of the
cross-sectional area of the counterplaten, while in older presses sliding
bar or wedge were in most cases arranged in front of the platen to simplify the casting. Presses with laterally moving die slides are again made
in this simpler design which even permits using a cast or forged, solid
steel plate as counterplaten. With frame construction presses ease of
access is ensured by arranging the delivery opening on one side of the
frame (see Fig. 151 and 81).

Fig. 151. Oil-hydraulic rod extrusion press in frame construction with inclined frame end permitting of gripping the extrusion close to the die (By Hydraulik, Duisburg)

Fig. 152. Counterplaten supported on stools on the base frame

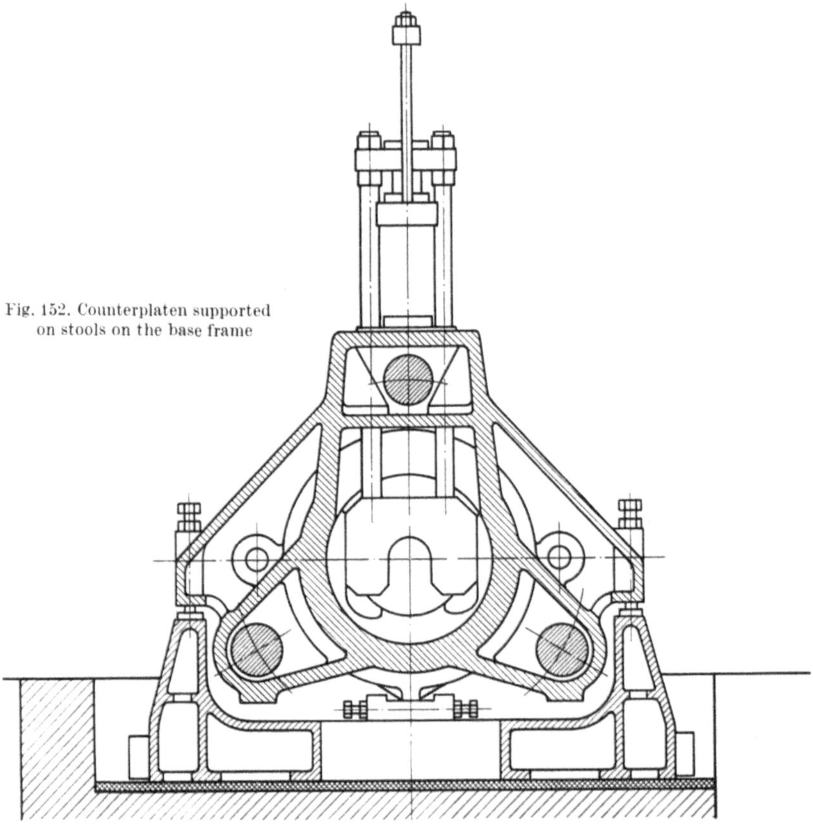

The deadweight of the counterplaten is supported on the base frame
on which it is arranged so as to be able to move in the longitudinal axis of

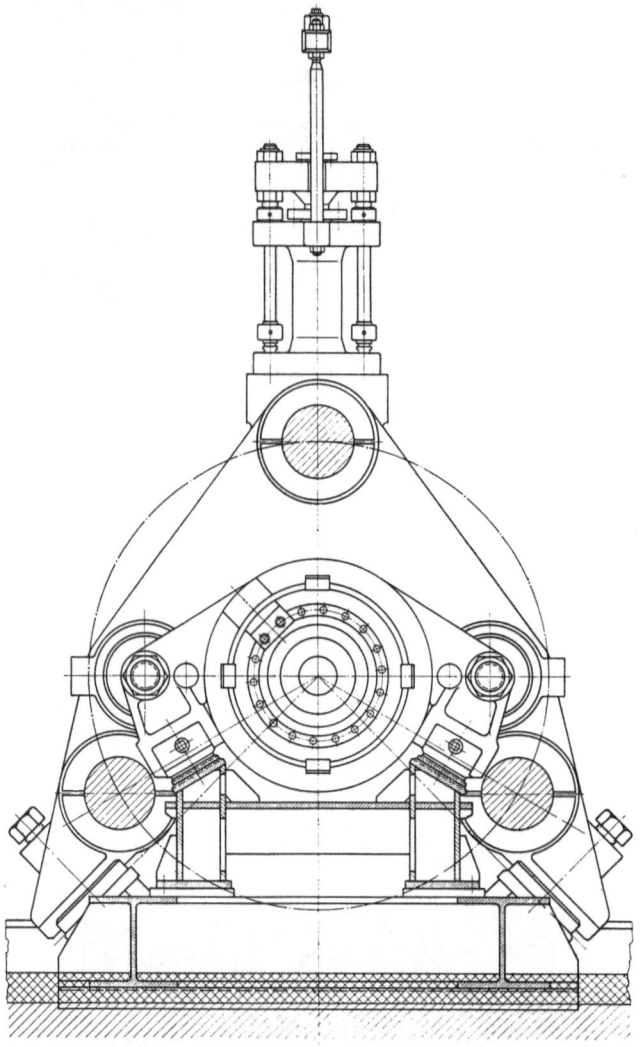

Fig. 153. Vee-shaped support of counterplaten on the base frame

the press. These motions are very small and are caused by the variation in
length of the press columns occurring during each extrusion cycle.

In continuous operation the counterplaten grows very hot and therefore the abutment must allow for equal distribution of thermal expansion

of the platen without offcentering. The simplest solution is to arrange the contact surfaces radially, as illustrated in Fig. 152 and 153. Fig. 152 shows the base frame with lateral supporting stools, on which the platen is supported through adjustable shoes. Lateral displacement is prevented

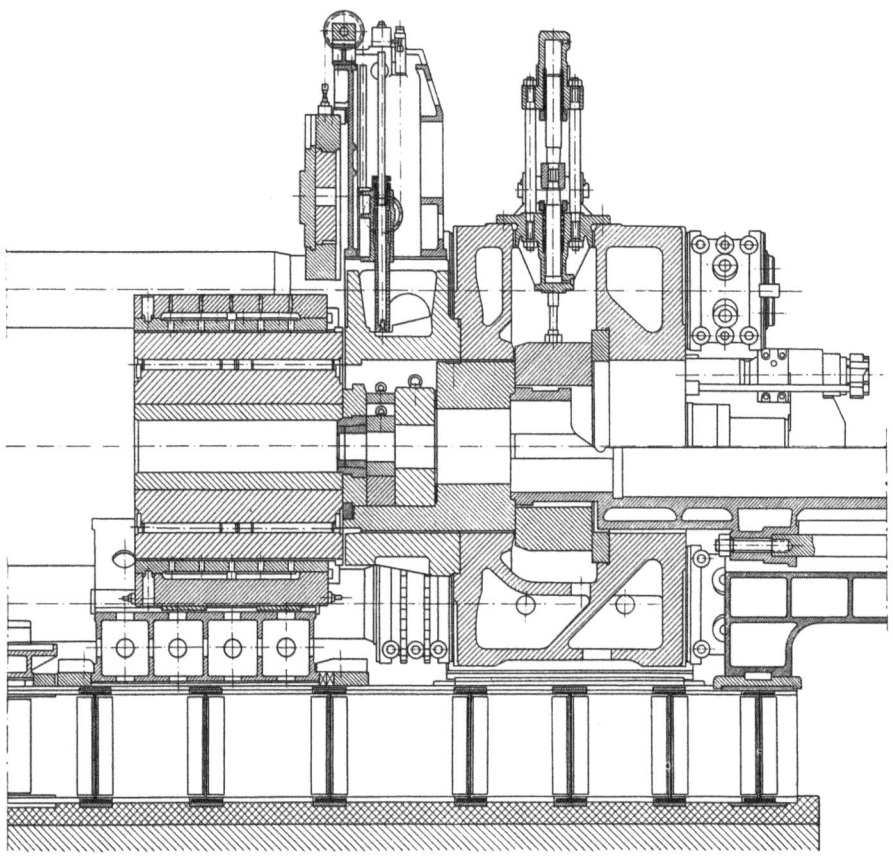

Fig. 154. Counterplaten with closure plate, open die-carrier and with opposite locking plates
(By Hydraulik, Duisburg)

by the bottom abutments. In Fig. 153 the counterplaten is supported by adjustable gibs on inclined, radially arranged surfaces on the base frame. The guide for the movable container-holder with the container supported in four keys, is designed in the same manner.

The counterplaten, illustrated in Fig. 154, is provided with a "blind die", i. e. a closure plate for the container, this being generally provided in heavy tube extrusion presses of capacities of 5,000 tons and more. It serves to prevent the wad of metal being forced through the die when piercing the billet in the container, and ensures that a thin disk of metal

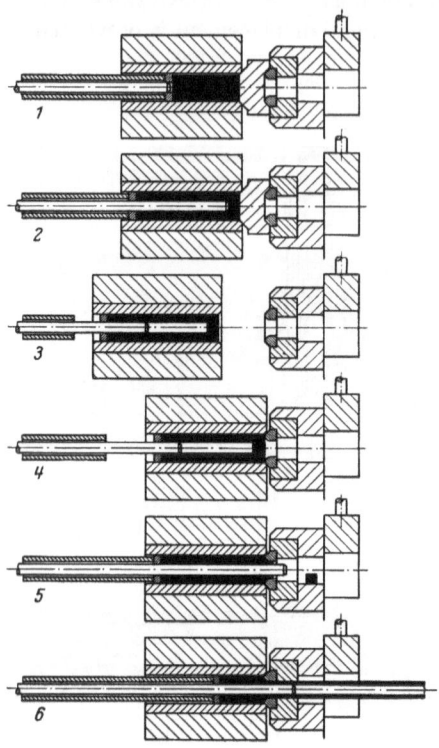

is left only which is ejected through the die after the closure plate has been lifted (Fig. 155). Thus the usable billet weight is considerably increased which results in better economy. The plate is suspended in a guide on two ropes which run over two stationary upper pulleys and are diverted over two loose pulleys. The loose pulleys are mounted on a slide which is solidly guided in the support and driven by a cylinder which moves over two stationary plungers. Lowering of the plate is effected by its own weight.

The sliding bar for locking the die-carrier is a two-part assembly. This design is preferred in heavy presses over the horseshoe shape because of the larger thrust carrying surface of the former. The two halves are positively moved against each other by levers. The drive unit consists of two opposed pistons, the cylinders of which are connected to each other through two columns. They serve at the same time as cover for the opening in the counterplaten (see Fig. 143).

Fig. 155. Sequence of operations in piercing against "blind die" and subsequent extrusion

m) Cylinder-Platens

In many rod and tube extrusion presses the main cylinder is combined with the pullback and piercer cylinders in one piece of steel casting to form a cylinder-platen which is provided with integrally cast lugs for the columns (see Fig. 88 and 89). The number of the auxiliary cylinders and their arrangement around the main cylinder have lead to a great variety in cylinder-platen designs.

In order to avoid heavy and complex castings, it is preferred – when heavy or frame type presses or a great number of auxiliary cylinders are concerned – to insert the cylinders separately in bores provided in the cylinder-platen, when, however, the additional attachment members and the less compact design of the press must be accepted (Fig. 156).

Fig. 156. Assembly of cast steel cylinder platen for inserted cylinders (By Hydraulik, Duisburg)

In many cases a composite construction is employed for reasons of design and casting technique, in which some of the auxiliary cylinders are cast integrally and some of them are inserted in integrally cast lugs (Fig. 67 and 73).

Long and narrow auxiliary cylinders are often forged or made from welded steel tubes.

For a long time the wall thickness of a cylinder was determined on the theory of elasticity, in which the tangential, radial and axial stresses are reduced in the direction of the principal stress, thus determining the

Table of Formulas 2. *Reduced stress in tangential direction.*
Formulas according to the theory of elasticity[1])

a) At internal pressure p_i

	Closed cylinder	Open cylinder
For any point in the cylinder wall	$\sigma_{t\,red} = -\,p_i\,\dfrac{1.3\,u_x^2 + 0.4}{u^2 - 1}$	$\sigma_{t\,red} = -\,p_i\,\dfrac{1.3\,u_x^2 + 0.7}{u^2 - 1}$
For the inner grain $(u_x = u)$	$\sigma_{t\,red} = -\,p_i\,\dfrac{1.3\,u^2 + 0.4}{u^2 - 1} \leqq \sigma_{z\,zul}$	$\sigma_{t\,red} = -\,p_i\,\dfrac{1.3\,u^2 + 0.7}{u^2 - 1} \leqq \sigma_{z\,zul}$
Required diameter ratio	$u \geqq \sqrt{\dfrac{\sigma_{z\,zul} - 0.4\,p_i}{\sigma_{z\,zul} + 1.3\,p_i}}$	$u \geqq \sqrt{\dfrac{\sigma_{z\,zul} - 0.7\,p_i}{\sigma_{z\,zul} + 1.3\,p_i}}$
Permissible internal pressure p_i	$-\,p_i \leqq \dfrac{\sigma_{z\,zul}}{1.3}$ gen.: $\;-\,p_i \leqq \dfrac{m}{m+1}\,\sigma_{z\,zul}$	$-\,p_i \leqq \dfrac{\sigma_{z\,zul}}{1.3}$ gen.: $\;-\,p_i \leqq \dfrac{m}{m+1}\,\sigma_{z\,zul}$

b) At external pressure p_a

	Closed cylinder	Open cylinder
For any point in the cylinder wall	$\sigma_{t\,red} = p_a\,\dfrac{1.3\,u_x^2 + 0.4\,u^2}{u^2 - 1}$	$\sigma_{t\,red} = p_a\,\dfrac{1.3\,u_x^2 + 0.7\,u^2}{u^2 - 1}$
For the inner grain $(u_x = u)$	$\sigma_{t\,red} = p_a\,\dfrac{1.7\,u^2}{u^2 - 1} \leqq \sigma_{d\,zul}$	$\sigma_{t\,red} = p_a\,\dfrac{2\,u^2}{u^2 - 1} \leqq \sigma_{d\,zul}$
Required diameter ratio	$u \geqq \sqrt{\dfrac{\sigma_{d\,zul}}{\sigma_{d\,zul} - 1.7\,p_a}}$	$u \geqq \sqrt{\dfrac{\sigma_{d\,zul}}{\sigma_{d\,zul} - 2\,p_a}}$
Permissible external pressure p_a	$p_a \leqq \dfrac{\sigma_{d\,zul}}{1.7}$ gen.: $\;p_a \leqq \dfrac{2\,m-1}{m}\,\sigma_{d\,zul}$	$p_a \leqq \dfrac{\sigma_{d\,zul}}{2}$ gen.: $\;p_a \leqq \dfrac{\sigma_{d\,zul}}{2}$

material stress in a given point which must be below the permissible stress:

$$\sigma_{t\,red} = \sigma_t - \frac{\sigma_r + \sigma_l}{m} \leqq \sigma_{z\,zul}\;{}^{2}).$$

σ_t, σ_r and σ_l are calculated after BACH's formulas (see p. 169) when additional stresses have to be considered which, for example, are caused by

---

[1]) See Table of Formulas 1, p. 169.
[2]) $\sigma_{z\,zul}$ = permissible pull stress.

Table of Formulas 3. *Comparison stress. Formulas according to the theory of deformation[1])*

a) At internal pressure p_i

	Closed cylinder	Open cylinder
For any point in the cylinder wall	$\sigma_v = -\, p_i \dfrac{\sqrt{3}\, u_x^2}{u^2 - 1}$	$\sigma_v = -\, p_i \dfrac{\sqrt{3\, u_x^4 + 1}}{u^2 - 1}$
For the inner grain $(u_x = u)$	$\sigma_v = -\, p_i \dfrac{\sqrt{3}\, u^2}{u^2 - 1} \leqq \sigma_{z\,\mathrm{zul}}$	$\sigma_v = -\, p_i \dfrac{\sqrt{3\, u^4 + 1}}{u^2 - 1} \leqq \sigma_{z\,\mathrm{zul}}$
Required diameter ratio	$u \geqq \sqrt{\dfrac{\sigma_{z\,\mathrm{zul}}}{\sigma_{z\,\mathrm{zul}} + p_i \sqrt{3}}}$	$u \geqq \sqrt{\dfrac{1.03\,\sigma_{z\,\mathrm{zul}}}{\sigma_{z\,\mathrm{zul}} + p_i \sqrt{3}}}$ approximate
Permissible internal pressure p_i	$-\, p_i \leqq \dfrac{\sigma_{z\,\mathrm{zul}}}{\sqrt{3}}$	$-\, p_i \leqq \dfrac{\sigma_{z\,\mathrm{zul}}}{\sqrt{3}}$

b) At external pressure p_a

	Closed cylinder	Open cylinder
For any point in the cylinder wall	$\sigma_v = p_a \dfrac{\sqrt{3}\, u_x^2}{u^2 - 1}$	$\sigma_v = p_a \dfrac{\sqrt{3\, u_x^4 + u^4}}{u^2 - 1}$
For the inner grain $(u_x = u)$	$\sigma_r = p_a \dfrac{\sqrt{3}\, u^2}{u^2 - 1} \leqq \sigma_{d\,\mathrm{zul}}$	$\sigma_v = p_a \dfrac{2\, u^2}{u^2 - 1} \leqq \sigma_{d\,\mathrm{zul}}$
Required diameter ratio	$u \geqq \sqrt{\dfrac{\sigma_{d\,\mathrm{zul}}}{\sigma_{d\,\mathrm{zul}} - p_a \sqrt{3}}}$	$u \geqq \sqrt{\dfrac{\sigma_{d\,\mathrm{zul}}}{\sigma_{d\,\mathrm{zul}} - 2\, p_a}}$
Permissible external pressure p_a	$p_a \leqq \dfrac{\sigma_{d\,\mathrm{zul}}}{\sqrt{3}}$	$p_a \leqq \dfrac{\sigma_{d\,\mathrm{zul}}}{2}$

the bending stresses emanating from the columns. With the help of Pois-son's ratio $m = 10/3$ the working formulas are found, which are tabulated in Table of Formula s 2.

Nowadays the theory of deformation is most commonly employed, the results of which correspond better with actual experiments. In this theory the stresses σ_t, σ_r and σ_l, calculated with the help of Bach's for-mulas, and additional stresses are reduced to the comparison stress

$$\sigma_v = \frac{1}{\sqrt{2}} \sqrt{(\sigma_t - \sigma_r)^2 + (\sigma_r - \sigma_l)^2 + (\sigma_l - \sigma_t)^2} \leqq \sigma_{z\,\mathrm{zul}}$$

[1]) See Table of Formulas 1, p. 169.

thus arriving at the formulas for the calculation of cylinders, tabulated in Table of Formulas 3.

Considering the repetitive stress sustained by the cylinder, the permissible stress σ_{zzul} has to be the fatigue strength of the material.

The cylinder bottom is given a heavily rounded and, if possible, a spherical shape[1]). For water-hydraulic systems, the cylinder is fitted with a bronze bush in which the operating plunger is guided, and with a radi-

Fig. 157. Combined 12,000-ton tube and rod extrusion press for working of light metals in four-column construction with three hydraulic cylinders arranged one above the other. Maximum billet diameter 800 mm, maximum billet length 1,800 mm. Courtesy of ALCOA, USA (By Schloemann, Düsseldorf)

ally adjustable stuffing box to provide for sealing means. Valuable empirical values for dimensioning of the volume of the stuffing box are tabulated in Table 23. Best suitable packing material is vulcanized cotton- or cellulose texture provided with a sealing lip with an adjoining chamber containing a grease-graphite mixture for the lubrication of the plunger.

In oil-hydraulic systems the plunger can be sealed by cast-iron piston rings, which retain their sealing qualities for years and have thus contributed to the wide-spread adoption of this drive system (see Table 23).

Heavy accumulator-driven presses, operating always at a constant pressure, are suitably provided with three cylinders, if their capacity amounts to more than 5,000 tons (see Fig. 157 and 158). In such case three different pressure stages – for which as a rule the ratio 1 : 2 : 3 is chosen – are attained by correspondingly arranging the press control. With equal diameters of plungers and cylinders pressure water is fed in the first stage to the middle cylinder only, in the second stage to the two outer cylinders only, and in the third stage to the three cylinders. The cylinders which are disconnected in pressure stages 1 and 2, are filled with low-pressure water during the plunger motion. If the first pressure

---

[1]) MÜLLER, E.: Hydraulische Schmiedepressen und Kraftwasseranlagen, 2. Aufl. Berlin/Göttingen/Heidelberg: Springer 1952.

stage is not so frequently employed, it may be advantageous to design the center working plunger for 2/3 of the press capacity and the two outer plungers for 1/3, so as to reduce the bending moments on the mov-

Fig. 158. Combined 8,000-ton tube and rod extrusion press for working of light metals in four-column construction with three hydraulic cylinders arranged side by side. Maximum billet diameter 500 mm, maximum billet length 1,200 mm. Courtesy of Aluminium-Walzwerke Singen (By Hydraulik, Duisburg)

able crosshead (see Fig. 158). The ratio of the plunger areas is then 1 : 4 : 1. A drawback in this solution is, however, that different seals and spare parts must be kept in stock for cylinders and filler valves.

With these three-cylinder presses the cylinders are inserted in the platen, as illustrated in Fig. 156; they are not subject to bending stresses and additional stresses. It is for this reason that steel casting continues to be used as cylinder material.

The cylinder-platen is rigidly bolted to the base frame and exactly aligned by means of heavy keys or links. When the press is suddenly released from pressure, the base frame is heavily loaded in shear and therefore a welded construction is recommended.

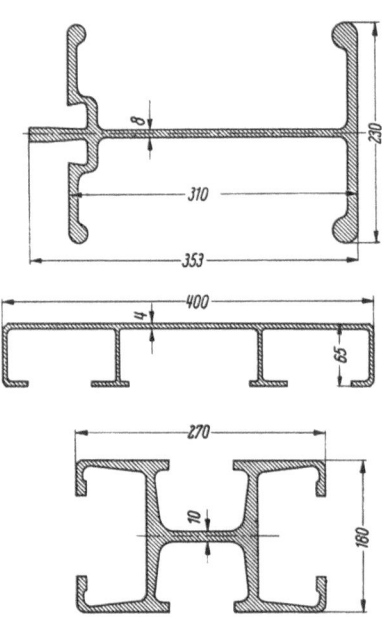

Fig. 159. Dimensions of light metal sections extruded on press shown in Fig. 158

13*

n) Columns

The positive connection of the cylinder-platen to the counterplaten is established through heavy columns made of open-hearth steel of high elongation and low notch sensitivity. One of the suitable steel grades is C 35. The column ends are in general secured in the two platens by split nuts and check nuts of cast steel, and prestressed. This pretension is obtained by overloading the press 10 to 15% while being started up, and tightening the inner nuts in this state. A small hand pump is in most cases used to increase the operating pressure. The pretension may also be obtained by heating the column ends which have been placed in the platens and then tightening the outer nuts by a value which has been evaluated mathematically. When column collars are used instead of inner nuts, this method must be applied, too. However, such columns cannot be adjusted in case a collar has been pounded into the platen due to the frequent load reversals.

Table 19. Saw-tooth threads for columns of hydraulic presses

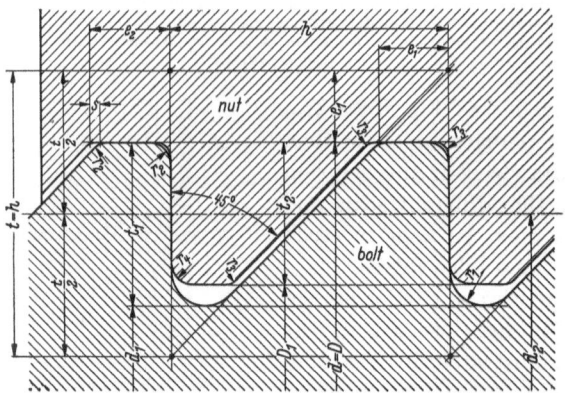

For pitch	Screw				Clearance	Nut			
h	t_1	e_1	r_1	r_2	s	t_2	e_2	r_3	r_4
5	2.875	1.25	0.619	0.375	0.26	2.5	1.51	0.25	0.375
6	3.45	1.5	0.742	0.45	0.28	3	1.78	0.3	0.45
8	4.6	2	0.990	0.6	0.32	4	2.32	0.4	0.6
10	5.75	2.5	1.237	0.75	0.36	5	2.86	0.5	0.75
12	6.9	3	1.485	0.9	0.40	6	3.40	0.6	0.9
16	9.2	4	1.980	1.2	0.48	8	4.48	0.8	1.2
20	11.5	5	2.475	1.5	0.56	10	5.56	1.0	1.5
24	13.8	6	2.970	1.8	0.64	12	6.64	1.2	1.8
32	18.4	8	3.960	2.4	0.8	16	8.8	1.6	2.4
40	23	10	4.950	3.0	0.96	20	10.96	2.0	3.0

Table 19 (Cont'd)

Screw			Pitch Diameter	Pitch	Nut	
Major Diameter	Core Diameter	Core Area			Major Diameter	Core Diameter
d	d_1	cm²	d_2	h	D	D_1
200	190.8	285.92	196	8	200	192
212	202.8	323.02	208	8	212	204
224	214.8	362.38	220	8	224	216
236	226.8	404.00	232	8	236	228
250	240.8	455.41	246	8	250	242
265	253.5	504.71	260	10	265	255
280	268.5	566.21	275	10	280	270
300	288.5	653.70	295	10	300	290
315	303.5	723.45	310	10	315	305
335	321.2	810.29	329	12	335	323
355	341.2	914.34	349	12	355	343
375	361.2	1,024.67	369	12	375	363
400	381.6	1,143.69	392	16	400	384
425	406.6	1,298.45	417	16	425	409
450	431.6	1,463.03	442	16	450	434
475	456.6	1,637.43	467	16	475	459
500	481.6	1,821.64	492	16	500	484
530	507	2,018.86	520	20	530	510
560	537	2,264.84	550	20	560	540
600	577	2,614.82	590	20	600	580
630	607	2,893.79	620	20	630	610
670	642.4	3,241.16	658	24	670	646
710	682.4	3,657.36	698	24	710	686
750	722.4	4,098.69	738	24	750	726
800	763.2	4,574.74	784	32	800	768
850	813.2	5,193.79	834	32	850	818
900	863.2	5,852.11	884	32	900	868
950	913.2	6,549.70	934	32	950	918
1,000	963.2	7,286.57	984	32	1,000	968
1,060	1,014	8,075.43	1,040	40	1,060	1,020
1,120	1,074	9,059.38	1,100	40	1,120	1,080
1,180	1,134	10,099.88	1,160	40	1,180	1,140
1,250	1,204	11,385.26	1,230	40	1,250	1,210

The column threads are of trapezoidal cross-section. Valuable dimensions and pitches are compiled in Table 19. Equal heights are chosen for nuts and check nuts $h = 1$ to $1.2\ d_1$, when d_1 = core diameter of column. The nuts are secured on the columns by flat bars to prevent slackening.

It is a well-known fact that it is the first pitches of a screw that take mainly part in the transfer of forces. Over these pitches the outside diameter of the column thread is therefore provided with a slight taper so as to distribute the load as far as possible over the other pitches, too.

The advantages and drawbacks of designing the presses in three- or four-column construction are realized best from a survey of their development.

The early rod extrusion presses were in four-column construction (see Fig. 57). The billet was inserted into the container from the rear, and container and ram could be easily mounted and dismounted through the

Fig. 160. Oil-hydraulic 1,200-ton extrusion press for light alloys in three-column design with crosshead being guided on the two top columns (By Hydropress, New York)

upper pair of columns with the help of a crane. Loading the billet through the counterplaten was very complicated and tiresome and proved to be particularly disadvantageous in the early tube extrusion presses with their increased billet output. Most of these presses were then designed as front-loading types. The billet was transferred from the furnace to the press by means of a pair of tongs suspended on a chain and shifted by hand on an overhead trolley. This billet transfer was obstructed by one of the top columns and thus the three-column design was chosen which ensures ease of access to container and ram, although it renders dismantling more difficult. Also, the top column is exposed to a higher degree to heat emanating from the container and therefore requires a special insulating protection.

With heavy presses of about 5,000-ton capacity and more it is appropriate not to transfer the billet into the press by means of an overhead trolley, but by a carriage. The three-column press is not suitable for this purpose, because the two bottom columns are relatively high for the rails. Preference is therefore given again to the four-column design which has the additional advantage of a better distribution of forces on the two platens, which results in lower bending stresses and lighter castings.

Recently the four-column design has again been employed for small and medium presses equipped with lateral die slides as this design facilitates the arrangement of the guide for the slide. It is further possible to arrange the shear for cutting the discard on the counterplaten (see Fig. 75). The three-column design is, however, more advantageous when the die is set in a rotating arm (Fig. 150).

Another interesting type of a three-column press is that with two columns arranged in the top horizontal plane on which the crosshead is guided, and with the shear being arranged on the counterplaten in connection with a lateral die slide. This kind of column layout is not exposed to fouling by foreign matter and does not require any particularly accurate adjustment as the press is not fitted for piercing. The bottom column lies in the horizontal central plane of the press and – on account of the guide for the crosshead being abolished on this column – permits of a simple feeding of the billets into the container (see Fig. 160).

Consequently, either design stands on justifiable grounds; however, instead of a two-column construction preference is given to the cheaper frame type press.

o) Movable Crossheads

Directly in front of the cylinder stuffing box the main plunger or piston is attached to a crosshead which carries the ram-holder with ram. In rod extrusion presses which serve solely for the manufacture of shapes, the crosshead is guided in its travel on the columns. For tube presses and also for rod presses used for the extrusion of hollow sections and tubes with floating mandrel, this type of guide is not sufficient as it is not accurate enough. With these presses the best solution is to arrange the guideway on the base frame thus eliminating the effects of temperature and elongation. The guideway consists as a rule of a cast-iron bed. The guide bars on the crosshead are suitably made of bronze and provided with adjusting wedges permitting of aligning the crosshead. This is necessary to ensure that the mandrel is held in co-axial alignment with the press and to keep to a minimum variations in the wall thickness of the tubes. In most cases the guideway takes a wide U-shape which permits of adjusting the crosshead both vertically and horizontally. The slideway is lubricated with oil or grease. Heavier presses are provided with a central lubrication.

The design of the movable crosshead is often influenced greatly by the type of piercer employed (see Fig. 88 and 89).

Chapter V

DRIVE SYSTEMS

All presses are driven by pressure-water or pressure-oil which is supplied either directly by pumps or by an accumulator installed between pumps and press. The choice of the drive system is a question of the particular requirements, economy of operation, and space available.

The pump output for the direct drive of a press must be rated so as to cover the maximum quantity required within a working cycle, viz. during the actual power stroke. Disregarding any losses, the maximum requirement is $N = \frac{Pv}{7,500}$ in H.P., when P = press capacity in kg and v = working speed of the main plunger in cm/sec. The required press capacity P varies over a wide range in accordance with the different working resistances of the alloys and cross-sections of the shapes to be extruded. The electric power input to the pumps being proportional at any instant to the required output, the direct pump drive is highly economical as far as power consumption is concerned. A comparison of the horsepower installed and the mean requirement during a full working cycle – which not only consists of the power stroke, but also includes return stroke, idle strokes and waiting times – shows that the less the proportion of the actual power stroke to the full cycle, the less is the efficiency factor of the available capacity. Although it is possible to assume a smaller motor capacity than that of the above equation, provided that the maximum of P is of short duration only and peak demands may be overcome by flywheel masses, the accumulation of energy thus obtainable is rather limited.

If the horsepower to be installed is to be rated for the mean requirement only, pressure fluid must be stored in an accumulator (see Fig. 161); the useful contents of the accumulator must equal the requirement during the power stroke, while being recharged throughout the cycle by the relatively small pump. The accumulator being constantly under approximately equal pressure, the pressure fluid is permanently supplied under the high pressure irrespective of the resistance met by the press, while the surplus hydraulic pressure has to be eliminated by throttling. The accumulator system will therefore not yield as economic a utilization of electric power as the direct drive system and therefore its range of application is limited.

From the foregoing it will be seen that the direct pump drive is to be preferred when the actual power stroke occupies the better part of the cycle, while accumulator drive is to be recommended when relatively high speed rates are called for and power strokes are of short duration.

Fig. 161. Pressure water accumulator for combined tube and rod extrusion press shown in Fig. 158

It may be said that the dividing line between the two types of drive, which of course is approximate, lies at a working speed of about 50 mm/sec.

With presses operating at slow speed rates, i. e. lead wire- and lead pipe presses, cable presses and rod extrusion presses, the working stroke is of relatively long duration – 10 to 20 minutes not being extraordinary – so that the accumulation of pressure fluid would make no sense. For tube extrusion presses, especially those working on heavy metals and at speed rates of up to 200 mm/sec, the accumulator drive will be preferred as the direct pump drive would be very expensive on account of the high pump capacity required.

There are still other points in favor of accumulator drive. One of its greatest advantages is that a number of presses can be operated simultaneously from one accumulator. If several presses operate in an extrusion shop or if a press, such as a tube extrusion press, is fitted with very many auxiliary drives, all of the presses and auxiliary cylinders may be connected to a common circuit which is fed from an accumulator. Thus space savings are considerable; also better supervision is ensured, especially when it is possible to install the accumulator station in a separate building outside the main shop. With direct driven presses, however, this problem is not so easily solved. When a press is driven by a pump to which another press which meets a somewhat smaller resistance, is connected through a control, the movement in the first press will stop immediately, because the pump pressure will adapt itself to the lowest working resistance. In direct pump operation it is therefore only possible to carry out successive motions. When the motions are to be performed simultaneously, one pump will have to be provided for each movement. Consequently, pumps in direct connection with a press are suitable for single drives only.

Another reason for the use of an accumulator drive even for presses working at slow rates is its shock-free operation. A pump plunger delivers pressure fluid at each revolution of the crankshaft in a manner similar to the shape of a sine curve. With the normal three-plunger pumps a wave-like delivery curve with tolerable irregularity, as shown in Fig. 163, is obtained. If, however, little leaks occur at the stuffing boxes or valves, these will affect considerably the shape of the curve and cause shocks in the piping in direct driven presses. As a consequence a color difference will occur on the surface of the shape extruded in a rod press, for example. In most cases this marking cannot be removed by subsequent drawing and will give rise to objections, if a very high surface finish is required.

Heavy cable presses for extrusion of aluminum sheathing (see p. 49) are operated by a combination of direct pump drive and accumulator drive. One or more pumps work directly into the main cylinder, whereas the pullback cylinders and the auxiliary drives are connected to an accumulator station with separate pump. The reason for this is the necessity of pulling back the ram at a high speed rate, this requiring a considerably higher power input than the power stroke proper. The latter is most suitably performed by direct pump drive, as the pressing- or delivery speed of the cable must be maintained accurately irrespective of the constant decrease in pressing power due to the reduction in the length of the billet. The constant delivery speed is ensured by the constant delivery volume of the pump, while in accumulator driven presses the speed rate changes in accordance with the varying working resistance.

During the past ten years the direct pump drive has found an extra-ordinarily wide application (see Fig. 162), this being due to the trend towards the use of pressure-oil drives instead of pressure-water drives. Pressure-oil pumps occupy only a fraction of the space required by pressure-water pumps (see Fig. 166) so that they may be used even when peak loads occur for which the accumulator drive would have been pre-

Fig. 162. Oil-hydraulic 1,400-ton extrusion press in four-column design (By Lake Erie, Buffalo, USA)

ferred formerly. The advantageous features of the direct oil drive with regard to the design of the presses have already been mentioned on p. 86. These include – to summarize them briefly – use of double-acting pistons, sealing by piston rings and elimination of corrosion hazards. Notwith-standing these advantages, pressure water pumps are still used – especi-ally when high pressures of 400 to 500 atm are required as the leakage losses of oil pumps would be extraordinarily high at such pressures (see p. 139). Water pumps are also preferred in cases which present potential fire hazards. Non-inflammable oils – although on the market already – are still very expensive.

Up to date the liquid mainly used for storing in hydraulic accumula-tors, is water, because accumulator stations are always connected with an extensive piping system which has to sustain shock pressures and vibrations often causing leaks at the flanged joints. Thus considerable quantities of water are lost; the water is drained through channels, soaked in the ground or it evaporates. If it were oil, it could not be removed in such a simple way and the losses would be enormous. It has further to be considered that oil must not come in direct contact with highly compressed air. Damage has repeatedly been caused by the ignition of oil vapors causing an explosion the reasons of which could not be ascertained. This has lead to the development of oil-accumulators, which are not loaded by compressed air, but by nitrogen.

a) Water-Hydraulic Pressure Pumps

There are two types of water-hydraulic pumps to be distinguished – the vertical and the horizontal pressure pump. The vertical type requires less space than the horizontal one and – fitted with overhead crankshaft – it is especially suitable for direct flat- or V-belt drive with a maximum capacity of 100 to 150 H.P. In general practice, however, separate gear drives mounted in a housing, are preferred over the belt-drives, although the initial cost of a gear drive is higher; they are difficult to connect to the high crankshaft, and nowadays vertical pumps are therefore seldom used.

If, however, the crankshaft of a vertical pump was arranged in the bottom – as in a vertical steam- or Diesel engine – and the working cylinder in the top, the stuffing boxes for the plungers would be over the crankshaft, due to which water might leak into the lower part of the crankcase and mix there with the oil for the lubrication of the bearings.

Best results have so far been obtained with horizontal pressure pumps in both single- and double-acting design. The double-acting type is equipped with a reversing linkage – with the exception of the few pumps that are provided with differential pistons – and has therefore found a relatively small range of application. The pump most frequently used, is of the single-acting three-plunger type, in which all components are easily accessible and which ensures a very uniform delivery of pressure water (see Fig. 163). Four-plunger pumps are hardly built because of their lower coefficient of uniformity, wheras one- or two-plunger pumps are more frequently used in the low capacity range up to about 10 H.P. Five-plunger pumps with improved coefficient of uniformity have only been built in vertical design. It is general practice to install at least two pumps so as not to lose any time in repairs. When particular value is attached to obtaining a delivery as uniform as possible, two three-plunger pumps with 30° staggered crankshafts may be coupled, thus obtaining a delivery diagram of a six-plunger pump, as illustrated in Fig. 163.

The design of a single-acting, horizontal three-plunger pressure pump is shown in Fig. 164. The base frame is made of cast iron. The crank drive is totally enclosed. The crankshaft runs in journal-bearings and is provided with either duplex or quadruple bearings. The duplex bearing ensures even bearing loads and offers the advantage of a reduction in overall width of the pump. A quadruple seated crankshaft is more favorably stressed; it has smaller bearings which nevertheless require a wider frame than the double-seated shaft. Both designs have proved under operating conditions.

The crosshead guides are of cylindrical or flat design and are in most cases provided with adjustable shoes. All of the sliding surfaces and bearings are to be equipped with forced-feed lubrication.

The pump bodies are designed with suction- and delivery valves being arranged either side by side or one above the other. The latter require one cover only and reduce the overall length of the pump. With the former inspection of the suction valves is easier, because the delivery valves do not have to be dismounted prior to inspection of the suction valves. Possibilities and limitations of the two types counterbalance each other. Further design details have fully been described in a previous volume[1]).

The pump is suitably driven by a squirrel-cage motor through a spur gear. The shafts are connected through flexible couplings to eliminate shocks on the work mains when starting the motor and to cater for small bearing

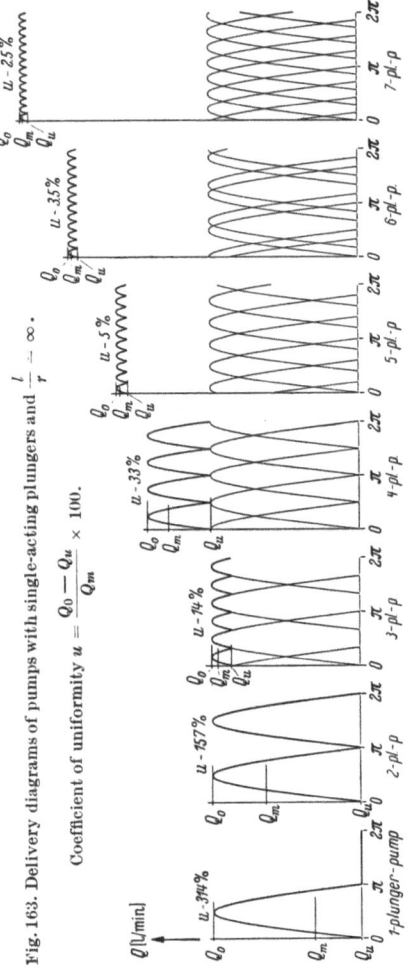

Fig. 163. Delivery diagrams of pumps with single-acting plungers and $\frac{l}{r} \div \infty$. Coefficient of uniformity $u = \frac{Q_0 - Q_u}{Q_m} \times 100$.

clearances. Table 20 shows a proposed standardization of the outputs of larger pump units as principally employed for extrusion presses. As a rule, the pump speed is in the region of 120 to 180 rpm at motor speeds of 1,000 to 1,500 rpm.

[1]) MÜLLER, E.: Hydraulische Schmiedepressen und Kraftwasseranlagen, 2. Aufl. Berlin/Göttingen/Heidelberg: Springer 1952.

The quantity of pressure water Q in litres/minute delivered by a single-acting multi-plunger pump is calculated with the help of the

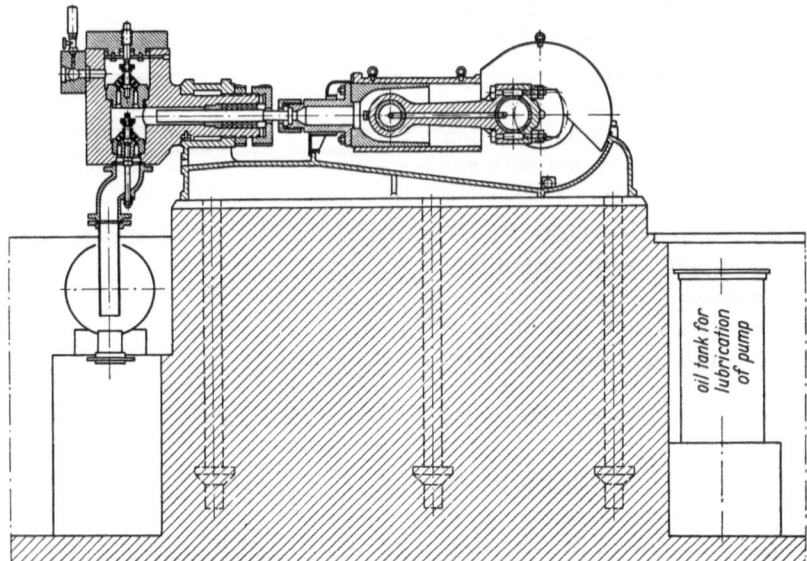

Fig. 164. Pressure water pump with valves arranged one above the other
(By Schloemann, Düsseldorf)

Table 20. Standard Dimensions of Pressure Pumps (Extract fr m DIN 2770). Power Requirements, Motor Capacities, Nominal Pressures, Deliveries

Power Required at the Crankshaft		Motor Capacity	Nominal Pressure in kg/cm²					
			100	160	200	250	315	400
H. P.	kW	kW	Delivery in litres/minute					
63	45	50	224	140	112	90	71	56
80	56	63	280	180	140	112	90	71
100	71	80	355	224	180	140	112	90
125	90	100	450	280	224	180	140	112
160	112	125	560	355	280	224	180	140
200	140	160	710	450	355	280	224	180
250	180	200	900	560	450	355	280	224
315	224	250	1,120	710	560	450	355	280
400	280	315	1,400	900	710	560	450	355
500	355	400	1,800	1,120	900	710	560	450
630	450	500	2,240	1,400	1,120	900	710	560
800	560	630	2,800	1,800	1,400	1,120	900	710
1,000	710	800		2,240	1,800	1,400	1,120	900
1,250	900	1,000		2,800	2,240	1,800	1,400	1,120
1,600	1,120	1,250		3,550	2,800	2,240	1,800	1,400
2,000	1,400	1,600		4,500	3,550	2,800	2,240	1,800

equation

$$Q = \frac{z f s n \, \eta}{1000}$$

where:

z = number of plungers
f = area of plunger in cm²
s = stroke of plunger in cm
n = number of revolutions per minute
η = volumetric efficiency, for which a mean value
η = 0.94 may be introduced.

The mean effective power N required at the crankshaft of a single-acting multi-plunger pump is in H.P.[1]

$$N = \frac{P_m \, v_m}{7{,}500 \, \eta_1}$$

when the mean total power P_m of z plungers, as referred to one revolution,

$$P_m = \frac{1}{2} \cdot z f p$$

is inserted in kg, when
p = hydraulic pressure in kg/cm²
$v_m = \frac{2 s n}{60}$ = mean speed of plunger in cm/sec
η_1 = mechanical and hydraulic efficiency, approx. 0.85.
Therefore

$$N = \frac{P_m \, v_m}{7{,}500 \, \eta_1} = \frac{z p f \times 2 s n}{2 \times 60 \times 7{,}500 \times 0.85} \, .$$

With

$$z f s n = \frac{1{,}000 \, Q}{\eta} = \frac{1{,}000 \, Q}{0.94}$$

hence

$$N = \frac{Q p}{360}$$

Q in litres/min
p in kg/cm²
N in H.P.

b) Oil-Hydraulic Pressure Pumps

It is possible without any difficulties to also employ oil as the transmission medium in the water-hydraulic pumps described in the previous chapter. The expenditure involved for these pumps would, however, be out of all proportion, as oil-hydraulic pumps are of much simpler and cheaper design, as there is no need to provide for stuffing boxes and separate systems for water and lubricating oil. Pump valves may also be omitted, if the cylinder bores are made to rotate over a cam plate.

---

[1] Calculation based on German horsepower (1 German H.P. = 0.9863 U.S. H.P.).

There are two types of oil-hydraulic pumps, viz. constant delivery
and variable delivery pumps. The output of the latter is controlled by
varying the stroke of the piston. The piston stroke chosen is very short
as compared with water pumps, thus rendering possible a considerable
increase in motor speed and to couple the pumps directly with the motor
(see Fig. 165). This type also permits of using an essentially larger number

Fig. 165. Oil-hydraulic 2,300-ton extrusion press in four-column design with pump platform arranged
above the main cylinder (By Lake Erie, Buffalo, USA)

of pistons. Not less than five plungers are generally used which results
in the delivery becoming practically perfectly smooth and free of shocks.
The increased number of pistons in turn results in an important increase
in efficiency and a more compact design of the pump which requires only
a fraction of the space of a water-hydraulic pump. Fig. 166 illustrates
the comparison of the space required by these two different pumps.

The stroke of constant delivery oil pumps is invariable and they are
thus of simpler construction than variable delivery pumps. A control
gear will, however, be required, if such pump is to operate an extrusion
press. In order to change the speeds of the press although the stroke of
the pump is invariable, part of the pressure oil may be by-passed through
a regulating valve into the return piping. When this change takes place
frequently and the oil grows too hot, it is more suitable to have the pump
driven by a variable-speed motor or to have the press driven by a number
of pumps, thus rendering possible to adjust the speed in steps.

Fig. 167, 168 and 169 show an oil-hydraulic pressure pump the outstanding feature of which being the provision of intake- and delivery
valves. This pump has proved itself especially at high pressures of 300
to 400 atmospheres on a continuous duty basis. Valve-less pumps have

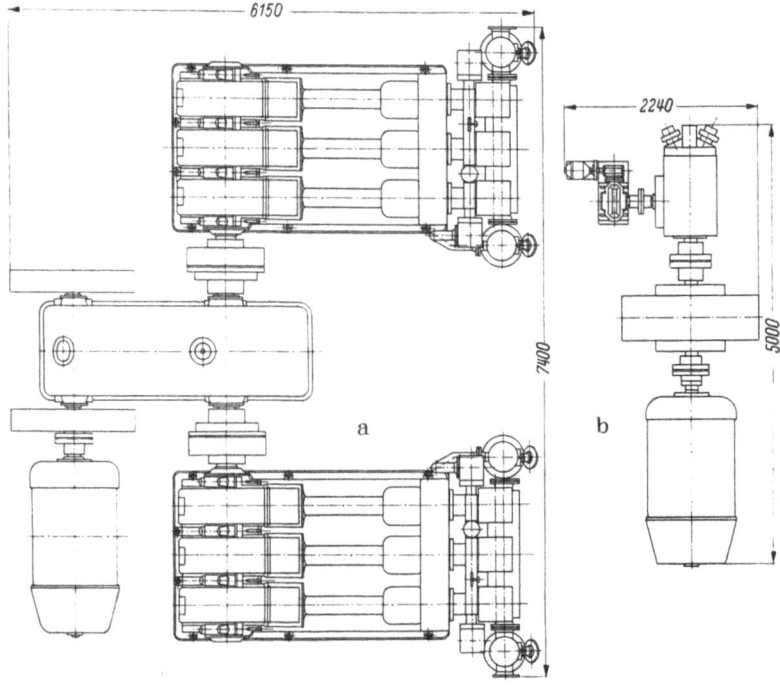

Fig. 166. Comparison of space required by (a) pressure water pump and (b) pressure oil pump of equal
capacity of approx. 2,000 l/min. at 200 atm.

not found a wide application for pressures of more than 250 atmospheres,
as in such case oil leakages at the cam plate would be excessive. The
cylinder block of the pump shown in Fig. 167 is a steel forging. The plungers, which are arranged straight in line, are pushed into the cylinder
bores through ball races by an overhead eccentric shaft having a constant stroke, and retracted during the suction stroke by springs. The intake and delivery valves are easily accessible after removal of a screw
fitting. The bores serve at the same time as valve seats and wear as in
water-hydraulic pressure pumps does not occur. The complete pump is in
an enclosed, oil-flooded housing.

Variable delivery oil-hydraulic pressure pumps are in axial or radial
design as illustrated in the schematic representation Fig. 170. The radial
pump possesses a stationary shaft a on which the pump body b, which is

directly driven by an electromotor, rotates. During each rotation the plungers c move. They run on rollers in an outer race d. When the centers of the pump body and the race are in one axis (see sketch a), the plungers stop in the position shown in the sketch, i. e. the delivery of the pump is

Fig. 167. Constant delivery pressure oil pump with suction- and delivery valves

Fig. 168. Section through pump shown in Fig. 167

zero. The race being shifted to the right (sketch b), the plungers perform a reciprocating motion on each revolution of the pump body. This motion increases as the race is shifted. Thus the delivery of the pump is regulable

Fig. 169. Twin pump; regulation of delivery by individual control of pumps (By Towler, England)

at random. To enable the oil to enter the cylinder bores on the intake stroke and to emerge from them on the discharge stroke, the stationary shaft a is provided with two ports e into which four longitudinal bores f

connect, of which two serve for the inlet and two for the outlet of the oil. The race being shifted into the opposite direction towards the left (sketch c), the motion is repeated with the only difference that intake- and delivery side of the cylinder change and that the flow of oil is reversed.

This characteristic feature of the pump is of extreme value when the deliveries of two mating bores are merged into one on the outside and hence piped to the respective power- and return sides of the press piston.

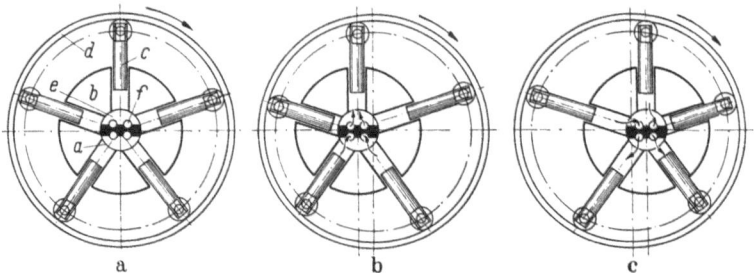

Fig. 170a–c. Schematic diagram of stroke variation of a radial piston pump

Thus the piston performs an advance and return motion and the control gear normally required with a constant delivery pump, can therefore be omitted. This simple circuit arrangement calls for equal-sized power- and return surfaces of the press piston which also results in equal speed rates in either direction. When these areas are of different size – as it is mostly the case in practice – and idling- and return strokes are to be performed at considerably higher speed rates than the power stroke, it is necessary to provide the press with a filling valve and a check valve. The corresponding circuit diagram is illustrated in Fig. 171.

During the idle stroke, as shown in sketch a, the oil flows from the supply tank at a pressure of 3 to 5 atmospheres through the filling valve which is mounted at the bottom of the cylinder, behind the main piston. The oil on the return side of the main piston flows to the intake side of the pump. The closed check valve in front of the supply tank prevents the oil from flowing back into the tank. The speed of the main piston depends on the intake capacity of the pump, i. e. on the intake quantity which is equal to the delivery quantity. For slowing down or stopping the operation the pump is swung back or brought into the mid-position respectively, i. e. set to "zero stroke". The idle stroke having been completed, the filling valve is closed by springs. The pump continues to deliver and builds up the operating pressure in the cylinder (sketch b). On the return side there occurs now a pressure drop, as the intake of the pump is higher than the quantity of oil flowing in from this side. There-

14*

fore additional oil passes through the check valve from the supply tank
on to the intake side of the pump. For the initiation of the return motion
of the press, intake- and delivery side of the pump are changed by swinging the pump housing into the opposite direction. The check valve is

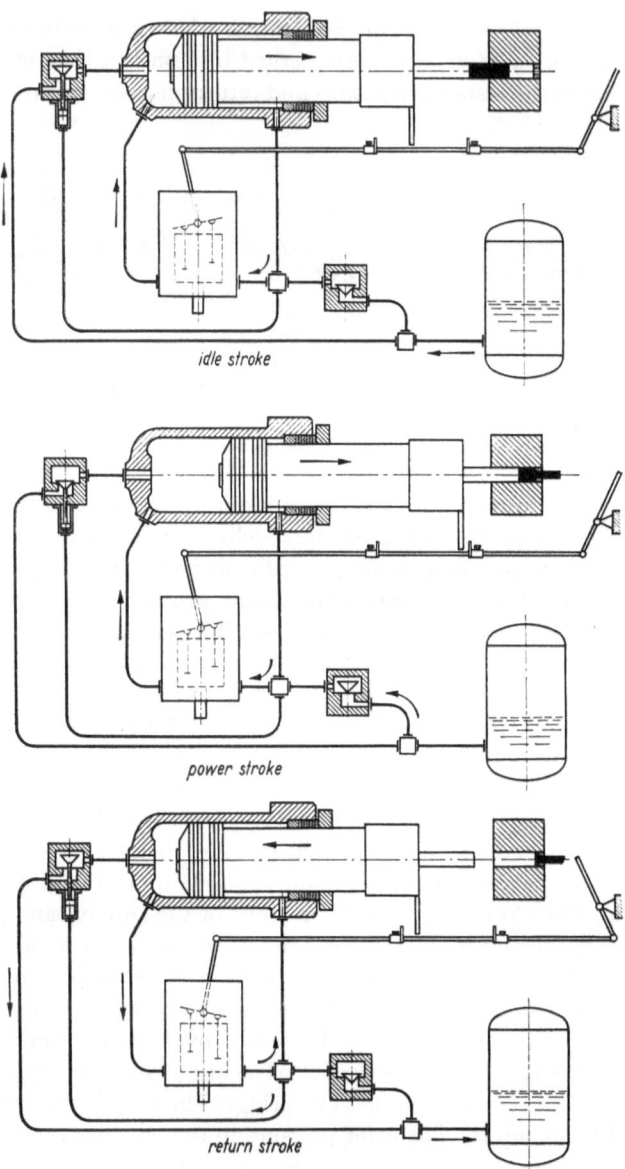

Fig. 171 a–c. Circuit diagram for direct pump drive with variable and reversible delivery of oil

thereby closed, while the filling valve is forced open (sketch c). The oil which is displaced by the piston, passes back into the supply tank with the exception of the quantity that is re-absorbed by the pump. To slow

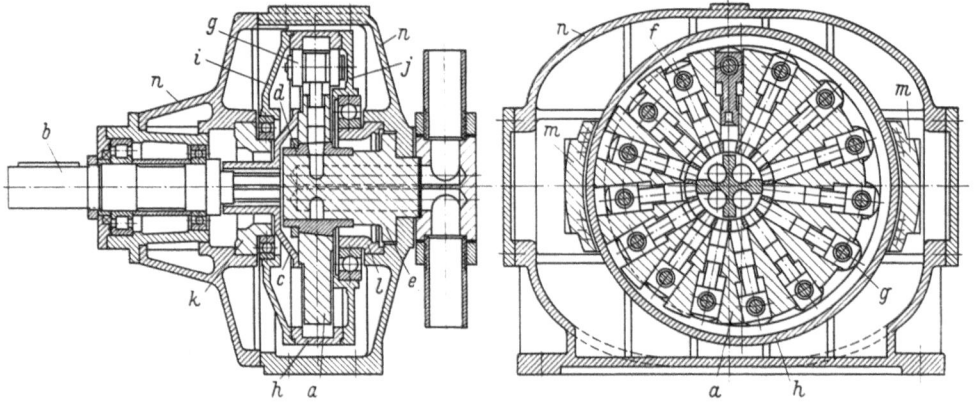

Fig. 172. Variable-delivery radial piston pump (By Kieler Howaldts-Werke, Kiel)

down the return speed or to stop the piston motion, the pump must be swung back or brought back to "zero stroke".

Fig. 172 and 173 illustrate a radial pump equipped with 15 plungers running in a rotating cylinder plate. The plate a is driven by the shaft b through the disk coupling c. On the opposite side the plate moves with a socket d over a dowel pin e which is provided with the control ports and the oil inlet and outlet bores. The bores are connected into a manifold with the two connecting pipings. To ensure solid guiding, the plungers are provided with an enlarged head having an axle f on which are mounted two rollers g on both sides of the cy-

Fig. 173. Shop photo of rotor of pump shown in Fig. 172

linder plate. The rollers lie against a race h which is supported on a cradle k, l by two plates i and j. This cradle k, l with the braces m moves in the split housing n thus controlling the stroke of the pump. The suction stroke motion is brought about by the centrifugal force of the plungers and rollers which come to rest against the race h.

A variable-delivery axial pump is shown in Fig. 174. The drive shaft a is seated in the pump housing in two bearings and has seven cylinder bores, arranged parallel to its axis, with pistons b. The pistons sit with spherical shoes c on a thrust bearing in the cradle d. At each rotation of

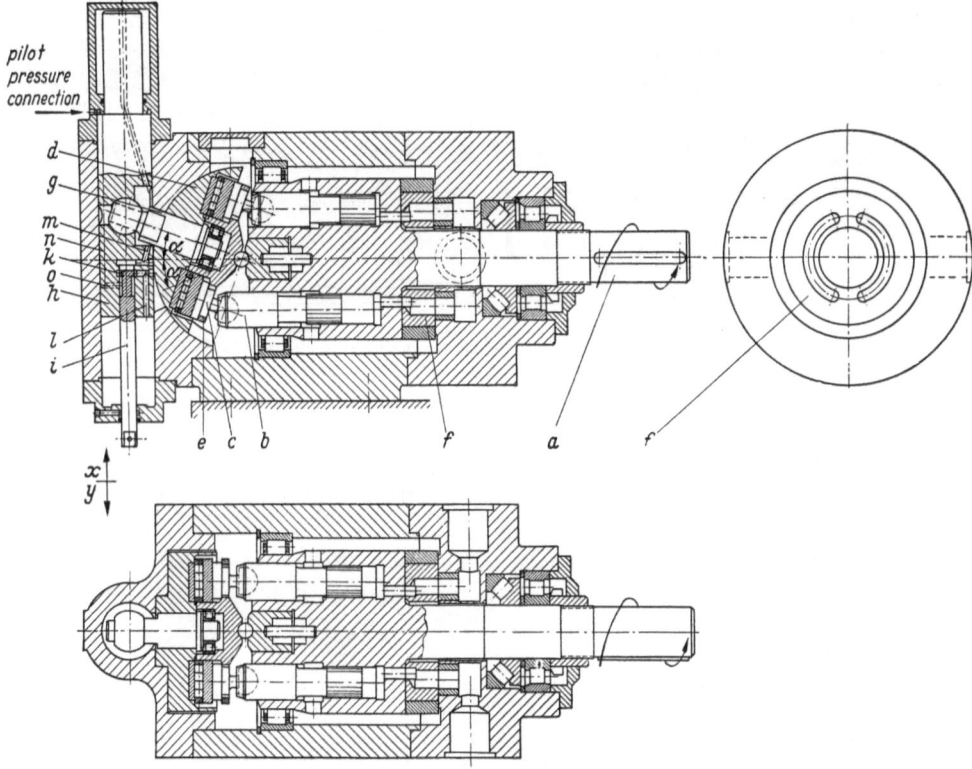

pilot
pressure
connection

Fig. 174. Variable-delivery axial piston pump

the shaft a the pistons b perform an intake and discharge stroke, the length of which is dependent on the degree of tilting of the cradle. The spherical shoes c are pressed against the thrust bearing by a driving disk e which is spring seated on the cylinder body. The stationary valve plate f is provided with two kidney-shaped ports which are connected to the cylinder bores and control the intake- and discharge cycle.

In pumps of up to about 50 H.P. the cradle d is swivelled by hand with the help of a pivot g; the angular displacement of the pivot is $x = \pm 15$ to $20°$ when the pump performs its longest stroke and the output attains its maximum value. In the mid-position stroke and output are zero. When passing over the mid-position intake- and discharge side change and the flow is reversed.

When human strength is no longer sufficient to adjust the cradle, a servo-motor with piston-valve h and slide-valve i is used. The servo-motor is driven by a small gear pump which produces a pilot pressure of about 20 atmospheres and is rated for 2 to 3 H.P.

When the slide-valve i is moved in direction y, its control edges k open the previously covered bores in the piston-valve h, thus enabling the oil under the piston-valve to pass through bores l and m into the chamber which is connected to the leak-oil circuit and arranged above the slide-valve. The piston-valve moves down and shifts the cradle until the bores in the piston-valve are covered again by the control edges. During this motion the pressure of the oil on the ring-shaped area of the piston-valve acts as the piston power. When the slide-valve moves in direction x, the pilot oil passes through groove n and bore o into the annular chamber and from there through bore l underneath the piston-valve h thus causing an upward movement which continues until the bores in the piston-valve are covered again by the control edges k. During this motion the constant pressure on the ring-shaped area is overcome by the piston power. The servo-motor thus acts perfectly as power amplifier.

In the stroke end positions of a press the variable-delivery pumps are automatically set to "zero stroke", when, however, a given oil pressure must still be maintained so as to hold the press piston against the filling pressure. The pump has to deliver a small quantity of oil which must correspond to the leakages in the complete system. The simplest means of attaining the two stop positions is to provide the control linkage with two lock nuts, as shown in Fig. 171, which are entrained by the main press piston and continue to swivel back the pump until the piston motion stops. The same effect is obtained when the linkage influences the servo-motor indirectly.

In order to attain the stop position after a partial stroke of the main piston, the linkage must be moved by the operator in a very sensitive manner. Less attention is, however, required for the operation of a valve- or slide-valve control, which is used with constant delivery pumps. In this case the control-lever has only to be shifted to the stop position whereby the feed line is closed and the motion of the main piston is stopped immediately. With variable-delivery pumps setting of a precise stop position is most difficult when it is to be attained in an automatic control cycle, i. e. without interfering manually. A movable stop is as a rule provided on the linkage which is rendered inoperative after the stopping period is completed and switches the pump back to delivery. These stops may be replaced by simple contacts in case of electrically

driven servo-motor slide-valves. As the motions are most easily controlled
with the help of a special control gear, not only variable-delivery pumps
are used for the operation of presses, but also pumps of the constant
delivery type.

Advantages and drawbacks of pumps of the radial and axial type are
open to debate, and there is a wide variety of both types available. In

Fig. 175. Variable-delivery axial piston pump; delivery up to 2,000 l/min. at 200 atm
(By Hydraulik, Duisburg)

general it may be said that the axial type is preferred for high pump capacities (see Fig. 175) on account of their lower circumferential speed and
better utilization of the material of the pump body.

c) Pressure Water Accumulators [1])

A hydraulic accumulator is supplied with pressure water whose economic pressure is in the region of about 200 atmospheres, by one or more
pumps (see p. 138). The complete station is suitably installed separate
from the presses in a room of its own.

The water operates in a closed circuit; it flows from the supply tank
to the pumps which feed it into the accumulator. The latter supplies it
through controls to the press and on completion of the power stroke the
water passes through the controls into a low-pressure air-vessel and
hence via an overflow valve back into the supply tank.

The two main types are the air-loaded and the dead-weight accumulator.

---

[1]) MÜLLER, E.: Hydraulische Schmiedepressen und Kraftwasseranlagen, 2. Aufl.
Berlin/Göttingen/Heidelberg: Springer 1952.

Dead-weight accumulators are obsolete and only very few have been constructed during the past 20 years. They consist of a vertically arranged cylinder whose piston is loaded by a heavy weight made up of cast-iron plates or a vessel filled with ore or scrap metal. The weight depends on the water pressure required; it amounts, for example, to about 140 tons at a piston diameter of 300 mm and an operating water pressure of 200 atm. The drawbacks of this accumulator type are its large overall

Fig. 176. Pressure water accumulator with two water bottles fed by three horizontal pumps
(By Hydraulik, Duisburg)

dimensions and the pressure surges liable to occur in it by the release of the kinetic energy of the heavy moving mass when the withdrawal of pressure water is suddenly stopped.

The air-loaded accumulator generally consists of a water bottle and a battery of air bottles connected to the former (Fig. 176). The ratio of the volume of water to the total volume of the bottles is as a rule about 1 : 10. The pressure drop between highest and lowest water level, i. e. on withdrawal of the total usable contents of pressure water, is about 10 to 12%. The compressed air serves the only purpose of loading the water to the required pressure and consequently, the air is not consumed.

Large accumulators are provided with a number of communicating water bottles, while in small installations the water- and air volume may be contained in a single bottle. Advantageous dimensions of bottles, which are most frequently of the seamless drawn type, have been published previously[1]).

[1]) Müller, E.: Hydraulische Schmiedepressen und Kraftwasseranlagen, 2. Aufl. Berlin/Göttingen/Heidelberg: Springer 1952.

There is no generally valid equation for the determination of the usable water contents V. In many cases $V = V_1 + V_2 + V_n$ is chosen, when V_1 through V_n are the volumes of water consumed by all of the presses which are supplied by the accumulator, in a simultaneous working cycle; the quantity of water delivered by the pumps during the same time, serves as reserve.

Having calculated the total pump output for the operation of a press installation, it is advantageous to split this output in such a way that the accumulator operates in conjunction with several – at least two – pumps. When there is not much pressure water consumed, only one pump may be used; in case of a breakdown of one pump, the operation is maintained by the other pumps.

Apart from accumulator and pumps a pressure water plant includes the control gear, fittings, and a high-pressure compressor.

The automatic control functions cause:

1. the pressure water delivery of the pumps to be stopped when the accumulator is filled;

2. the starting of the delivery after withdrawal of a given quantity of water from the accumulator;

3. the closing of the piping when the accumulator is empty to prevent compressed air being carried into it;

4. the indication of the water level.

There are many proved control gears for these various purposes. The method of control is based either on the water level irrespective of the internal pressure of the accumulator, or on the pressure drop occurring at the respective water levels. For extrusion presses the former method is preferred, because the latter provides no trustworthy indication of the water level in case of unforeseen or unnoticed losses of air.

In all of the accumulators now in use there are no control elements, such as floats, pistons etc. arranged in the water bottle, as troubles have been experienced in such designs.

The principles of working of one of the many accumulator controls[1]) may be followed with the help of Fig. 177. In this well-known unit the valves are electrically controlled and the impulses are made by the different water levels.

The control consists of two communicating bores containing mercury. One of the bores is connected to the air chamber of the accumulator and the other to the water chamber.

Rise and fall in the water level in the accumulator cause the mercury level in the air chamber to change in the same manner, when the travels

---

[1]) MÜLLER, E.: Hydraulische Schmiedepressen und Kraftwasseranlagen, 2. Aufl. Berlin/Göttingen/Heidelberg: Springer 1952.

of the two liquids are inversely proportional to their specific gravities. Contact pins arranged in the air chamber of the control gear body, touch the level and close low-voltage circuits, thereby energizing solenoids via

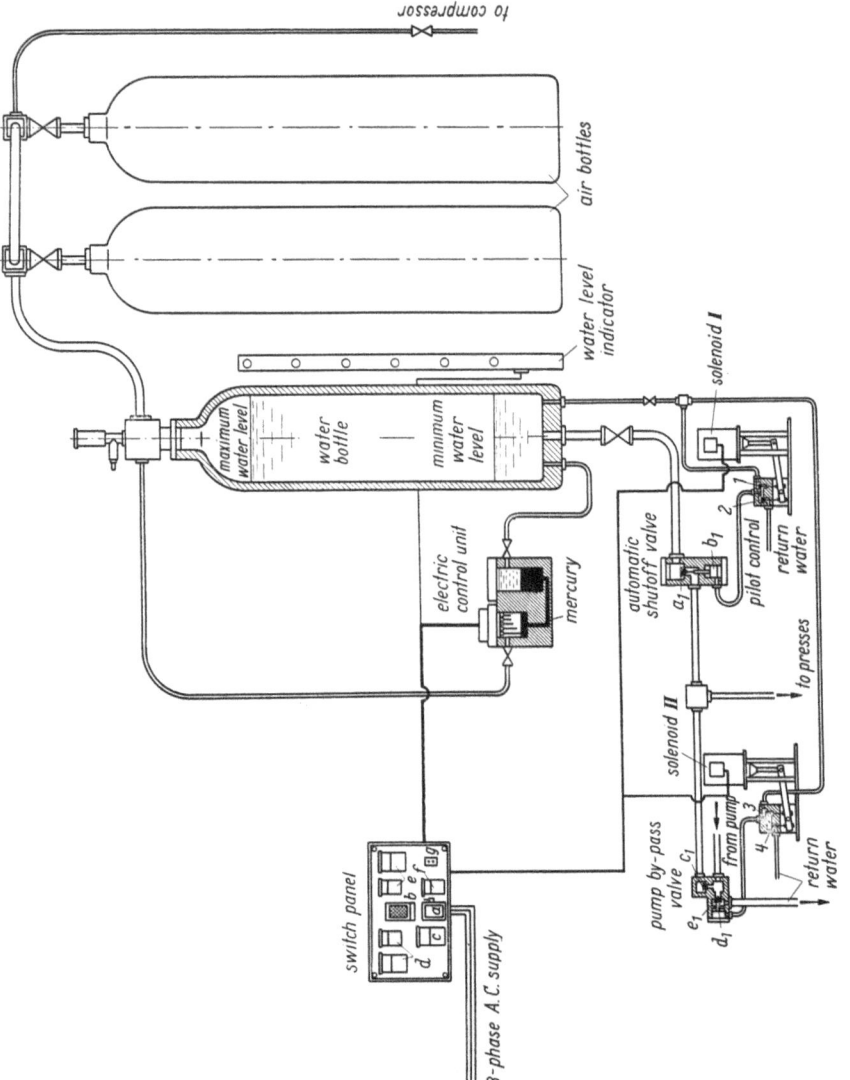

Fig. 177. Circuit diagram for pressure water station controlled by water level

relays. These solenoids in turn operate hydraulic pilot controls for the main valves, which cannot be opened or closed directly by the solenoids as the lifting capacity of the latter is insufficient.

In the position drawn in Fig. 177, the longest contact pin touches the mercury level. When the withdrawal of water continues, the last circuit in the air chamber will be broken due to the fall in the mercury level and due to the release of solenoid I the pilot control for the automatic shutoff valve is changed over. Now a connection is established between the cylinder bore b_1 under the lifting piston for shutoff valve a_1 with the return piping through the opened valve 2 of the pilot control. The shutoff valve is closed and withdrawal of water from the accumulator, i. e. any passing of compressed air into the piping, is thus prevented.

With the water level in the accumulator rising again due to the delivery of the pumps, when the pressure water flows through the check valve c_1 and the shutoff valve a_1 which also acts as a check valve, the proportionally rising mercury level causes the pilot control to change over by energizing solenoid I when the last but one circuit is closed. Valves 1 and 2 are opened and closed respectively. Pressure water is passed to cylinder bore b_1; the lifting piston opens the shutoff valve a_1 and re-establishes the connection between accumulator and press. The electric contact system being so arranged that the shutoff valve is reversed or closed only when the circuit for the longest contact pin is cut, there is a time interval between the individual reversing operations due to a so-called "cycling stroke" taking place in the accumulator; a high frequency of operations which would be especially inconvenient when the water- or mercury level is unsteady, is thus avoided.

The mercury level continuing to rise up to the last contact pin which corresponds to the highest water level in the accumulator, solenoid II on the pilot control for the pump by-pass valve is deenergized. Solenoid II drops; valves 3 and 4 are closed and opened respectively; cylinder bore d_1 behind the lifting piston for by-pass valve e_1 is connected to the return piping. The pumps are set to idling and the water delivered by them, is pushed through the opened by-pass valve e_1 and the return piping back into the supply tank. Check valve c_1 prevents the pressure water from passing out of the accumulator during idling.

Re-starting of the pumps or closing of the by-pass valve by reversing the pilot control takes place, when the last but one contact pin is cleared, i. e. after a "cycling stroke" has been performed as already explained in the operation of the shutoff valve.

On account of the great frequency of operations required it is not possible to stop the pump motors when the highest water level has been attained in the accumulator instead of changing over to by-pass. Disengageable couplings for the gears are not employed either, as the hydraulic by-pass devices are more efficient and less expensive.

The water level is indicated in the accumulator room and on the press control desks by lamps which are connected to the low-voltage circuits of the solenoids.

Faults in the cycle of control operations due to current failure are eliminated by the so-called "closed circuit current principle", i. e. when the solenoids are deenergized, the automatic shutoff valve closes and the pump by-pass valve opens.

The electrical operation of the control units permits of placing the accumulator at any distance from the pumps. In plants having a wide hydraulic mains system, it is also possible to install a number of accumulators at various locations and place the pumps in a common room. This arrangement ensures small cross-sections of pipes; furthermore shocks and heavy pressure drops when several presses work simultaneously, are eliminated.

The fittings required for an air-loaded accumulator include a hand-operated shutoff valve arranged directly on the pressure-water connection, hand-operated stop valves for the air bottles, a safety device which stops the pump motor when a given pressure is exceeded, as well as a control panel with relays and appertaining pushbutton switches for the solenoids, a circuit breaker, a transformer, the indicator lamps for water level and electric current.

One or more small multi-stage high-pressure compressors which are directly driven by motors of about 5 to 25 H.P., are used to pressurize the accumulator. The initial filling occupies a few days, whereupon the compressor will have to be used from time to time only to replenish air when losses due to small leaks or to absorption have occurred.

Chapter VI

HYDRAULIC CONTROLS

Designing and calculating a control is preceded by the preparation of a hydraulic circuit diagram, based on the rule that one inlet- and one outlet valve each or one double-acting slide-valve each is required for one reciprocal motion of the plunger. Consequently, a four-valve control is to be provided for the advance and return motion or the up-and-down motion of the piston of a horizontal or vertical press respectively. Examples of this are the controls for the press and the die-carrier shifting device, illustrated in Fig. 92.

A two-valve control also answers the purpose when, for example, the pullback cylinders of an auxiliary device of the press are connected directly to the pressure conduit. In such case the main piston, while advancing, has to overcome the constantly acting pullback power and must therefore have correspondingly larger piston- and valve cross-sections. The same applies if the pullback power is established by means of a weight or springs. Such simplified control with two valves is recommended for various auxiliary motions under full pressure and low piston forces, such as in shears, container shifting devices, wedge-type sealing devices, etc., as shown in Fig. 92. It would be less suitable, however, for the die-carrier shifting device, because the controlled cylinder and the piston would have to be doubled in size. A pullback device working under constant pressure, may by no means be employed for motions comprising one idle and one power stroke, such as all motions of the main piston. The reason for this being that the main cylinder is filled with low-pressure water or oil during the idle stroke, the pressure of which is not sufficient to overcome the constant pullback power.

All modern extrusion presses are equipped with controls which permit of prefilling the main cylinder so that on the one hand pressure-water or oil is consumed only for the power strokes proper and, on the other hand, relatively high speed rates are attained during the idle strokes. The so-called filling valve being nothing but a large water- or oil discharge valve which is balanced by a small pilot valve and which is pushed open by a drive piston the cylinder of which is connected to the pullback piping.

If, however, high idle speeds and a slow pullback speed are required, the drive piston may be omitted, provided that the cross-section of the controlled release valve is sufficient and prefilling may be carried out by a simple check valve. Design and calculation of the filling valves have already been described in one of my previous books[1]) so that another detailed explanation should not be necessary.

a) Hand-Lever Controls

Fig. 178 illustrates a control consisting of a forged block with incorporated valves or slide-valves. Slide-valves are employed for oil-drive only, since they may be ground so as to be tight; for pressure-water operation they would have to be sealed by packings whose service life would be rather short on account of the high pressures required in extrusion presses. The control housing is fixed on a trestle in which is set the drive shaft for the opening and closing of the valves or slide-valves. The valve stems or slide-valve rods are connected to the drive shaft through joints. Controls with cam plate drive have found less application as the stems tend to seize easily.

The valves may be used for both pressure-water and pressure-oil. There is quite a variety of types on the market which have also been described in detail already[1]). All valves are balanced so that they may

---

[1]) MÜLLER, E.: Hydraulische Schmiedepressen und Kraftwasseranlagen, 2. Aufl. Berlin/Göttingen/Heidelberg: Springer 1952.

easily be opened against pressure. Pressure inlet valves which are opened
by means of a simple stem, are, as illustrated in Fig. 178, provided with
a balance pin which serves to equilibrate the control shaft. If this was

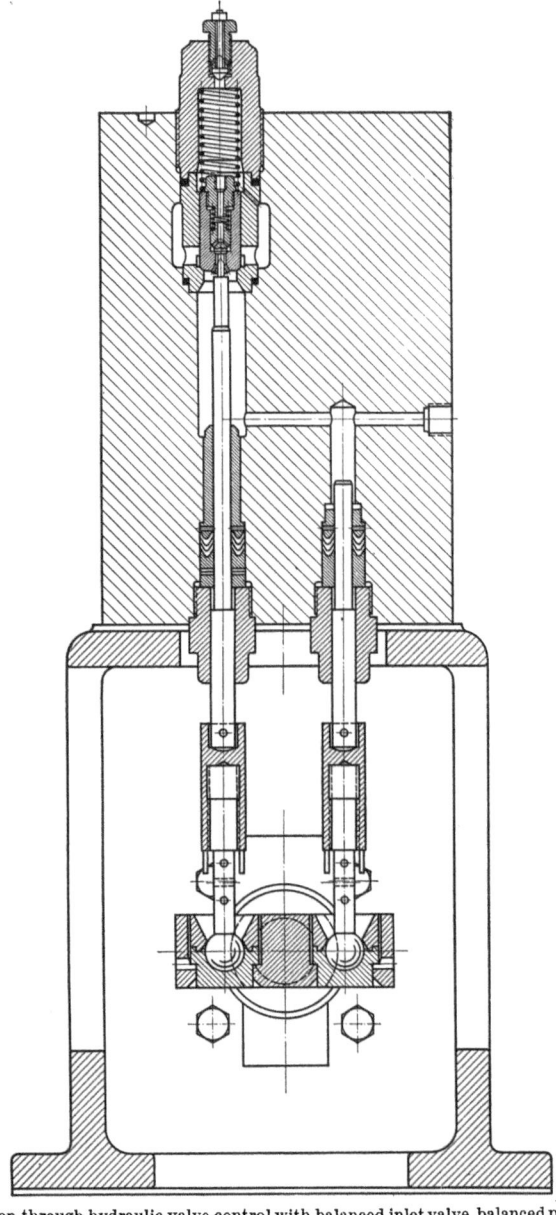

Fig. 178. Section through hydraulic valve control with balanced inlet valve, balanced pin and balanced
lever shaft

not provided, the force of the stem after opening an inlet valve, would cause a torsional moment on the control shaft which would impede the operation of the control. A balance pin is, however, not required in valves equipped with a through stem which is sealed in the upper valve cap and is thus perfectly balanced.

The free valve cross-section f depends on the velocity v and the cross-sectional area F of the main plunger, as well as on the rate of flow V of the fluid in the valve seat.

Therefore

$$F v = f V$$

and

$$f = F \frac{v}{V} .$$

Velocities chosen for extrusion presses:

$v = 0.3$ to 0.4 m/sec for idle- and pullback motions and
$v = 0.1$ to 0.15 m/sec for power motions if the delivery speed of the material extruded does not call for any restriction.

The maximum flow rate shall not exceed
$V = 30$ to 40 m/sec for inlet valves at operating pressures $p = 200$ to 315 atm. on account of wear of valve seats, pressure drop occurring and the temperature rise of the fluid. For outlet valves, which are loaded with a considerably lower pressure during the pullback motions, are chosen
$V = 10$ to 15 m/sec at $p = 10$ to 20 atm.

Cross-sections of filling valves are calculated as a rule for a rate of flow of
$V = 3$ to 7.5 m/sec at $p = 2$ to 5 atm.

The pressure drop h caused by the rate of flow V, can approximately be determined from the equation $V = \psi \sqrt{2\,g\,h}$ when the flow factor $\psi = 0.25$. In order to minimize losses and to eliminate shocks, the flow rates in the piping are kept considerably lower at
$V = 8$ to 10 m/sec and $V = 3$ to 6 m/sec for pressure- and return pipes respectively.

The control shaft is rotated with the help of a hand-lever which engages with a latch in a number of registers on a graduated arc. The various positions of the hand-lever are given by the valve-lift diagram; the maximum angular displacement is about $75°$.

The larger the cross-section of the valves the more increase the resistance met by the stem and the power required to actuate the hand-lever. With heavy presses it is therefore impossible for the operator to turn the control shaft directly with the help of a hand-lever. In such cases the con-

trol is actuated by a servo-motor operating in a similar manner as the
servo-motor for the pump described on p. 215; a schematic drawing is
illustrated in Fig. 179. When the hand-lever a is moved in the direction
towards the dash-dotted end position, rod b pulls on the one-arm lever c
which pivots in d and transmits the motion to rod e. Thus the valve
lever f is pivoted around g, outlet valve 2 is opened, and the stem under

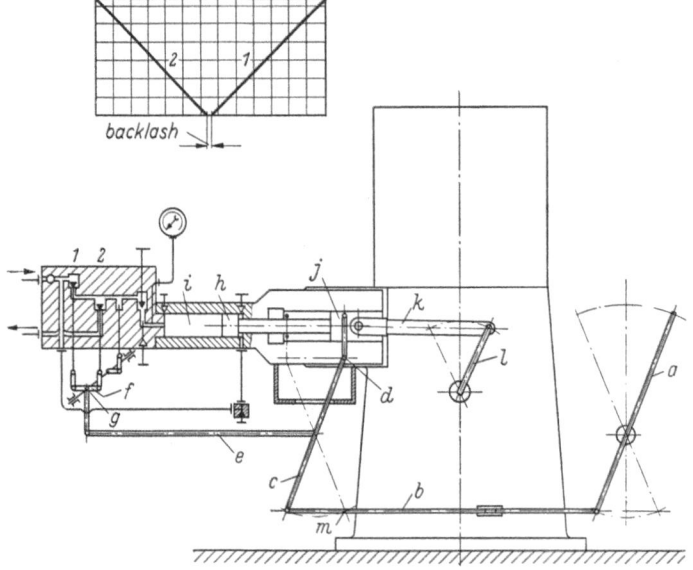

Fig. 179. Schematic drawing of servo-motor drive for a hydraulic valve control

the closed valve 1 is pulled down further. The servo-motor piston h
moves to the left due to the constant force on its annular area, while the
fluid escapes from the cylinder i through valve 2. The piston motion is
transmitted to the crosshead j and from there via connecting rod k and
lever l to the control shaft. Along with the crosshead j the center of
rotation d shifts to the left, too; lever c has now its center of rotation in
m, the rod e moves back, valve 2 is closed, and the piston h stops. When
moving the hand-lever a back into its initial position, the motions are
repeated in the reverse order, when inlet valve 1 is opened and the pressure in the cylinder i overcomes the constant back pressure on the annular area. The piston speed may be adjusted by the throttle valve so
that the described motions do not take place one after the other, but
almost simultaneously. The control shaft responds immediately to the
motions of the hand-lever and the position of the latter is always in
correspondence with a given position of the shaft. Consequently, the
servo-motor serves only to step up the force of the hand-lever.

b) Motor-Driven Controls for Remote-Controlled
and Automatic Operation

The various hand-lever controls for tube and rod extrusion presses
are arranged in such a way that the operator can closely watch the ram
motions and the charging operations. A drawback to this system is the
large space required and the distances between the various hand-levers,
on account of which the operator has to change his position frequently

Fig. 180. Underground arrangement of the control of a 2,500-ton tube and rod extrusion press

during each extrusion cycle. Furthermore, the numerous pipings branching off the controls obstruct the access of the press.

It is for these reasons that – already many years ago – the controls
were installed underground (Fig. 180), that the hand-levers were arranged
closely to each other and that these were connected to the control shafts
via a mechanical or hydraulic linkage. This development and the demand
for automation of the working operations soon led to electric remote
drives (Fig. 181) being actuated by pushbuttons or small control levers
from a central desk. These drives operate directly or via a servo-motor,
as already described, the control shaft.

The simplest type of electric remote control is the end-position change-over control which shifts the control shaft either to the pressure- or return position, when, however, the main piston cannot be stopped nor is it possible to regulate its speed. Such motions occur in hydraulic shears and saws, die shifting devices, wedge-type locks etc. The control shaft is advantageously driven by a 3-phase a.c. motor through a

Fig. 181. Remote-control desk of a tube and rod extrusion press (By Hydraulik, Duisburg)

small enclosed gear (see Fig. 182). The motor capacity is about 0.5 kW as referred to a duty factor of 60%. Two pushbuttons are provided on the control desk to set – via relays – the motor to clock- or counterclockwise motion. Stopping is automatically ensured by the control shaft through limit switches. The motor is suitably provided with a slide armature brake to render impossible any over running.

In oil-hydraulic presses which are controlled by balanced piston-valves, there is only a small force required and therefore a solenoid in conjunction with a spring may be used. As a rule, the solenoid is deenergized in the return position of the main piston. Large control slide-valves are moved by a piston, when the solenoid operated piston-valve serves as a pilot control.

The main control of tube and rod extrusion presses – whether accumulator driven or directly driven by variable-delivery oil-hydraulic pressure pumps – has to meet much higher demands. They must be able to regulate the working speed of the press at least in steps, and set the plunger motions to *advance, stop,* and *return.* This could be attained by arranging a number of intermediate switches on the control shaft, which, however, is not made use of, as there are better and simpler methods.

Fig. 182. Hydraulic control with electrically driven control shaft

A well-known type of remote-controlling the control shaft or the gudgeon of a variable-delivery pump, which has found a wide application, is that with a 3-phase a.c. motor working in conjunction with a gear and an electric compensating control. In this design two potentiometers are opposed to each other in the form of a bridge connection, viz. the *control potentiometer* and the *compensation potentiometer.* The former is mounted in the control desk of the press and may be scanned by a hand-lever which shifts a contact. The compensation potentiometer measures the angular position of the control shaft from where all the motions are transmitted to the potentiometer by a contact lever.

As long as the potentiometer bridge remains in electrical equilibrium, the electric drive of the control shaft has no reason to perform a rotation. However, as soon as the hand-lever of the control potentiometer is moved in any direction thus disturbing the equilibrium, the potential difference of the potentiometer bridge releases an electric control pulse, which is amplified by electronic tubes and transmitted to the motor contactors. Through a gear the drive motor rotates the control shaft in the desired direction, while the contact lever which scans the compensation potentiometer, joins in this motion. Thus an equilibrium is re-established on the electric bridge arrangement; the control pulse transmitted via the tubes to the contactors extinguishes and the motor is deenergized. The angle of adjustment of the potentiometer lever on the control desk is therefore proportional to the angle of rotation of the control shaft.

The over running of a motor after switching off – even when minimum values, such as occurring in a slide armature brake motor, are concerned – affects the accuracy of the control which for this reason cannot be employed for a stepless and very sensitive speed control of a press. When using commercial 3-phase a.c. brake motors of max. 3 to 4 kW, sufficient accuracy of control at high adjusting speeds may be attained by the electronic amplifier system even with largest control units.

In tube and rod extrusion presses, stepless remote-controls are employed almost exclusively for the regulation of the pressing speed. In presses which are driven by a hydraulic accumulator with a constantly operative and rather uniform pressure potential, regulation is done by means of a throttle gate valve or a throttle valve. The speed of directly driven oil-hydraulic presses is controlled by the infinite regulation of the delivery of the pumps.

When this remote control is to be directly driven by electric motors, a d.c. shunt-wound motor may be used to drive the infinitely regulable gear to which the control shaft or the gudgeon of the pump is connected. The switching principle is the same as with the 3-phase a.c. motor, viz. an electric bridge consisting of *control-* and *compensation potentiometer*. Week pulses result in a low number of revolutions, strong pulses in a high motor speed. Consequently, there is a proportionality – to a certain degree – between the angle of adjustment and the speed of adjustment, whereby the over running – such as occurring with the 3-phase a.c. motor and being so detrimental to the stepless regulation – is eliminated.

This equipment being very expensive, preference is now given to a servo-motor – similar to that employed with the pump in Fig. 174 – the piston of which is pilot controlled by a solenoid operated slide-valve. The servo-motor is driven by a small gear pump which delivers the pressure oil for a 30 to 40 atm control circuit.

The solenoid operated slide-valve concerned is of a special type, the so-called moving-coil regulator. Contrary to the former with its invariable stroke, the latter can perform from its mid-position, which is determined by a coil-spring-arrangement, variable strokes towards the plus or minus side. The stroke length of the moving coil, which, being coupled directly, operates the small pilot control slide of very low inertia, is proportional to the electric control current. This means that in case of high values being measured by the electric bridge, a large flow of oil is released which will cause the servo-motor piston to perform its correspondingly long stroke very quickly. As soon as the measured values decrease when approaching the electrical equilibrium, the stroke of the moving coil decreases and the speed of the oil flow deaccelerates, whereby the speed of the piston is slowed down and the piston travels into the required position in a very sensitive manner.

The electric remote controls do not only have to transmit control pulses from the control desk to the press. In many cases they have to perform regulating operations, such as, for example, maintaining a predetermined pressing speed or adjusting it positively. When using a stepless electric remote control which works on the electric bridge principle (see Fig. 183), the electric measuring bridge may be arranged in such a manner that the measured values of an electric speed measuring device are compared by the compensation potentiometer. In such case the control potentiometer on the control desk is disengaged.

Another operation to be performed is to stop the main ram at any desired position by servo-controlling the high-pressure pumps. In this the zero-stroke adjustment of the pump is most difficult (see p. 215), because the slightest inaccuracy in the remote control would cause the pump to deliver a small quantity to the plus or minus side, thereby moving forward or backward the main ram, which may then, for example, damage the billet loader in the center of the press. In such case the abnormal ram motion caused by any source of error whatever, is used as the correction

Fig. 183. Stepless electric remote control of a combined tube and rod extrusion press

means in that a third control potentiometer is added to the measuring bridge consisting of the two potentiometers. This third potentiometer is coupled to the ram motion in a predetermined stroke position and causes the oil pump to pivot in the opposite direction. Should this correction have been too intense, i. e. the pump having been set to the other delivery side, the motion is repeated in the opposite sense, when the amplitudes of this cycling motion will drop to zero very rapidly. This regulation is so precise that even in case of hardly noticeable aberrations from the predetermined stop position the ram is forced to adjust to the position at rest.

In a tube or rod extrusion press the cycle of operations can be performed semi- or fully-automatically through the electric remote control in a very reliable manner without any complicated apparatus. Such regularly repeating cycle consists of a series of motions of the main ram and of all of the auxiliary devices. By arranging electric limit switches in the press it is possible to obtain an automatic sequence switching without having to interfere with the control desk, which will be safe in operation, as a motion can only be initiated when the previous one has been completed. The control pulse to initiate such a sequence of motions is in most cases given by actuating a start button on the control desk. An electric master switch permits of stopping the automatic cycle at any time and of returning the pulse to the various impulse makers in the control desk. Thus the press operator may interfere at any time in the event of any faults occurring or if operations other than the normal ones have to be performed by the press.

Chapter VII

PACKINGS, PIPINGS, VESSELS, AND FITTINGS

Packings are used for stationary and sliding components. The packing points must be serviced very carefully, as constant leaks will always entail lengthy repairs. Packing material to be used depends on the pressure to be sustained by the packing and the kind of liquid used. Extrusion presses are mostly operated by water or light oil. The water is suitably mixed with water-soluble oil which serves to prevent corrosion and to lubricate the plungers. During operation the temperature of the liquid should not exceed 50 to 60 °C; although such temperature will not have any impairing effects, it is advisable to keep the temperature lower so as to protect the packings.

The pipes of heavier extrusion presses are laid in tunnels being covered by slabs of reinforced concrete or checkered plate. The tunnels should be of sufficient depth to allow walking space for an easy inspection. In case of space being limited or in order to avoid the high expenditure involved, tunnels of smaller clear dimensions may be provided and manholes arranged at branching points only. These manholes would accomodate the flanged joints, whereas the pipes in the rest of the circuit would be welded. In the tunnels the pipes have to be solidly anchored at 4 to 5 m spacings so as to prevent pipe ruptures due to vibrations which may be caused by shocks of the pressure medium due to inclusions of air or sudden deacceleration of the head of liquid. Structural irons are set in the concrete, on which the pipes are fixed on wooden fixtures by means of clamping collars.

The vessels, to which the circuit is connected, include the air-vessels of the presses and the open storage tank of the pumps from which the water or oil is supplied to the pumps and into which the liquid is passed back.

Fittings employed include manifolds, shutoff devices, pressure gauges, venting- and draining devices for the liquid.

a) Packings for Stationary Components (Flanges etc.)

Two types of packings are employed, i. e. packings that are deformed when installing them, and self-acting packings installed in packing chambers. The former include the metallic seals consisting in the main of copper, lead and soft iron, as well as soft packings of rubber, Klingerit etc. Sealing is brought about by the material being clamped and deformed between two sealing surfaces, when the unevenness of the surfaces to be sealed will be equalized by the material. Disk-shaped sealing rings of rubber laminated with canvas to increase the strength, would tear open and burst at high internal pressures. This is the reason why they are used for low pressures only such as occurring, for example, in suction-, filling-, drain- or return pipes. In flanged connections, as per Table 21, the packings are accomodated between two shoulders within the screw disk which facilitates installation and removal.

Pressure pipes with high operating pressures are sealed by copper rings. The flanges are provided with a recess which corresponds to the outside diameter of the sealing, and a corresponding shoulder (see Table 22) to accomodate these rings. The rings are thus centered and secured against bursting. The copper sealing ring must be endless and is cut off a pipe. Rings that are rolled from copper strip and welded at the joint, are unsuitable. Due to the fact that copper does not corrode in water or oil, a ring after having been disassembled, may well be used again; should it have been deformed heavily, it will be appropriate to anneal it prior to re-installation. Liquids – such as mercury – which

Table 21. Dimensions of Pipes and Flanges for Pressures up to 10 atm
(DIN 2632)

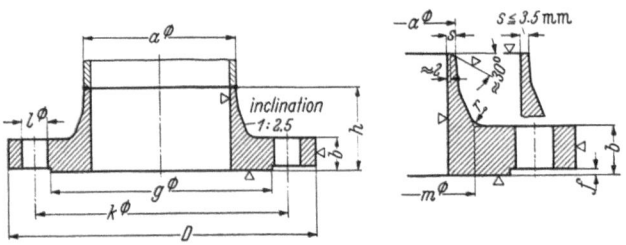

	Pipe	Flange				Neck			Raised Face		Bolts		
Nom. Diam.	Outside Diam.	Diam.	Thickness	Diam. of Bolt Circle	Height	Diam.	Thickness	Radius	Diam.	Height	Number	Thread	Diam. of Hole
NW	a	D	b	k	h	m	s	r	g	f		inch	l
15	18	95	14	65	35	30	2.5	6	45	2	4	$^1/_2$	15
20	25	105	16	75	38	38	2.5	8	58	2	4	$^1/_2$	15
25	30	115	16	85	38	42	2.5	8	68	2	4	$^1/_2$	15
32	38	140	16	100	40	52	2.5	8	78	2	4	$^5/_8$	18
40	44.5	150	16	110	42	60	2.5	8	88	3	4	$^5/_8$	18
50	57	165	18	125	45	72	3	8	102	3	4	$^5/_8$	18
70	76	185	18	145	45	90	3	8	122	3	4	$^5/_8$	18
80	89	200	20	160	50	105	3.5	8	138	3	4	$^5/_8$	18
100	108	220	20	180	52	125	4	8	158	3	8	$^5/_8$	18
125	133	250	22	210	55	150	4	10	188	3	8	$^5/_8$	18
150	159	285	22	240	55	175	4.5	10	212	3	8	$^3/_4$	22
200	216	340	24	295	62	232	6.5	10	268	3	8	$^3/_4$	22
250	267	395	26	350	68	285	6.5	10	320	3	12	$^3/_4$	22
300	318	445	26	400	68	335	7	10	370	4	12	$^3/_4$	22

Tables 21 and 22 reproduced by courtesy of Deutscher Normenausschuß.

destroy copper sealings, are sealed by soft iron gaskets. Their shapes and dimensions are the same as those of copper rings.

Metallic seals are exposed to the liquid on the sealing surfaces, whereas soft packings must be placed in special packing chambers in case of high operating pressures and kept away from the stream of liquid. Installation dimensions can be kept very closely and they are particularly suitable for sealing of screwed joints, covers, valve seats, screw plugs etc. Metallic seals are mainly preferred for flanged joints. Depending on the materials and shapes used, the soft packings may be subdivided in mechanically sealing, self-sealing and hydraulically sealing rings.

The mechanically sealing rings comprise all rubber rings which are made in various cross-sectional shapes. Round rings – also called 0-rings – are placed in rectangular, triangular or similarly shaped grooves and must be prepressed 10 to 15% of their thickness, while the cross-section

Table 22. Dimensions of Flanged Fittings
(Extract from DIN 2798 through 2800)

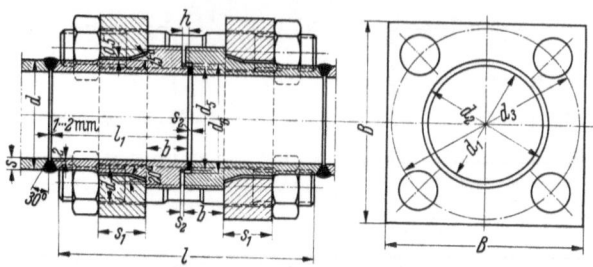

Operating Water Pressure	Nom. Diam. NW	Pipe DIN 9871		Flange DIN 2807					Welding Collar DIN 2810			Sealing Ring			Bolts DIN 2813	
		d	s	B	s_1	d_1	d_2	d_3	d_4	l_1	b	d_i	d_5	s_2	d_7	l
200 atm	10	18	4.5	70	20	22	28	60	33	60	17	9.5	15	2.5	M 16	110
	15	25	5.5	70	20	29	35	60	40	60	17	15	20.5		M 16	110
	20	30	6	80	20	34	40	72	46	70	23	19	25.5		M 16	120
	25	35	6	80	20	39	46	72	53	70	25	24	30.5		M 16	120
	32	44.5	7	100	25	50	58	90	65	85	25	33	39	3	M 20	140
	40	57	8	120	30	61	70	110	78	90	27	42	51		M 24	160
	50	63.5	8	150	35	68	79	135	85	105	30	49	58		M 30	190
	65	89	10	170	40	94	105	160	115	110	30	71	82		M 30	200
	80	108	12	200	50	114	127	190	140	140	41	87	102	4	M 36×3	260
	100	133	16	240	60	141	158	230	170	170	53	105	120		M 42×3	300
	125	159	18	280	70	167	186	270	205	200	61	129	144		M 48×3	360
315 atm	10	18	4.5	70	20	22	28	60	33	60	17	9.5	15	2.5	M 16	110
	15	25	5.5	70	20	29	35	60	40	60	17	15	20.5		M 16	110
	20	30	6	80	20	34	40	72	46	70	23	19	25.5		M 16	120
	25	35	6	100	25	39	47	90	53	80	25	24	30.5		M 20	140
	32	51	9	120	30	57	66	110	73	90	27	34.5	43.5	3	M 24	160
	40	57	10	150	35	63	74	135	83	100	30	39	48		M 30	190
	50	76	12	170	40	83	94	160	105	110	30	50	61		M 30	190
	65	108	18	200	50	116	129	190	142	140	41	73	86		M 36	260
	80	133	24	240	60	144	161	230	175	170	53	87	102	4	M 42×3	300
	100	159	28	280	70	171	190	270	206	200	61	105	120		M 48×3	360
	125	191	32	330	80	206	227	325	245	230	70	129	144		M 56×4	400
400 atm	10	18	4.5	70	20	22	28	60	33	60	17	9.5	15	2.5	M 16	110
	15	35	10	80	20	40	46	72	54	70	23	16	22.5		M 16	120
	20	44.5	12	100	25	50	58	90	66	85	25	24	30.5		M 20	140
	25	51	12	120	30	57	66	110	74	90	27	28	35	3	M 24	160
	32	57	12	150	35	63	74	135	84	100	30	34.5	43.5		M 30	190
	40	63.5	12	150	40	70	81	135	90	110	30	42	51		M 30	190
	50	76	14	180	50	83	96	160	110	140	41	54	65		M 36×3	260
	65	108	20	220	60	117	134	205	150	170	53	73	86	4	M 42×3	300
	80	133	26	260	70	145	164	245	180	200	61	85	100		M 48×3	360
	100	159	30	310	80	172	193	295	210	230	70	105	120		M 56×4	400
	125	191	35	350	90	207	230	340	250	250	73	129	144		M 64×4	440

Table 22 (cont'd.)

	200 atm	315 atm	400 atm
$d_6 = d_5 + 0.5$ mm	up to NW 25	up to NW 25	up to NW 20
$d_6 = d_5 + 1$ mm	from NW 32	from NW 32	from NW 25
$h\ \ = 5$ mm	up to NW 25	up to NW 25	up to NW 20
$h\ \ = 6$ mm	for NW 32 to 65	for NW 32 to 50	for NW 25 to 50
$h\ \ = 8$ mm	for NW 80 to 125	for NW 65 to 125	for NW 65 to 125

of the groove must be about 10 to 30% larger than that of the 0-ring depending on the respective hardness of the ring used. At operating pressure the ring is forced against the joint of the groove.

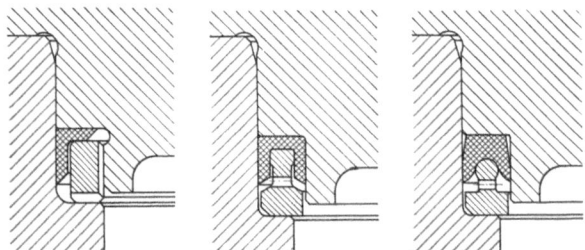

Fig. 184. Types of packing collars

Vulcanized fiber rings are of the self-sealing type. Due to the absorption of liquid the ring swells and presses against the chamber walls, thus ensuring a perfect sealing (see Fig. 178). Depending on the liquid employed, the clearance of the ring in the chamber is about 5 to 10% of its cross-sectional area. For dimensioning of the packing chamber exact data must be obtained regarding the swelling capacity, which, for example, is much lower in oil than in water.

Hydraulically sealing packings include all packing collars to the inside of which is admitted the hydraulic pressure, which causes the sealing lip to deflect and to be pressed firmly against the surface to be sealed. Packing collars are made of leather and synthetic materials and their shape varies with the intended use, operating pressures and liquids. Some designs are illustrated in Fig. 184.

b) Packings for Sliding Components (Pistons etc.)

Highest demands have to be met by such packings. They have to seal well pistons of presses and pumps in the stuffing boxes and to dissipate the friction heat quickly to the outside. Dimensioning of stuffing boxes and choice of suitable packings have already been dealt with (see

Table 23. Dimensions of Packing Spaces of Stuffing Boxes
(Dimensions in mm)

Diam. d mm	Width b mm	Pressure p atm up to	Soft and Lip Packings Depth of Space h	Height of Packing / Lip Packings h₁	Number of Rings	1 × U-shape Depth of Space h	Height of Packing h₁	Centering z	3 × V-shape up to 200 atm / 4 × V-shape over 200 atm for Leather Packing Depth of Space h	Height of Packing h₁	Height of Packing prior to Swelling	Centering z
10	6	100	38	30	3	26	20	3.6	38	32	30	3.6
		315	48	40	4							
		630	60	50	5							
25	7.5	100	48	36	3	32	25	4	45	38	35	4
		315	64	48	4							
		630	80	60	5							
80	10	100	60	48	3	40	30	6	55	44	40	7
		315	80	64	4							
		630	100	80	5							
125	12.5	100	75	60	3	50	35	10	70	55	50	10
		315	100	80	4							
		630	125	100	5							
225	15	100	90	75	3	55	40	9	80	60	55	14
		315	120	100	4							
		630	150	125	5							
450	20	100	120	96	3	75	55	12	100	78	70	14
		315	160	128	4							
		630	200	160	5							
1,000	25	100	150	120	3	90	65	15	120	95	85	15
		315	200	160	4							
		630	250	200	5							
over 1,000	30	100	180	150	3	110	80	18	140	110	100	18
		315	240	200	4							
		630	300	250	5							

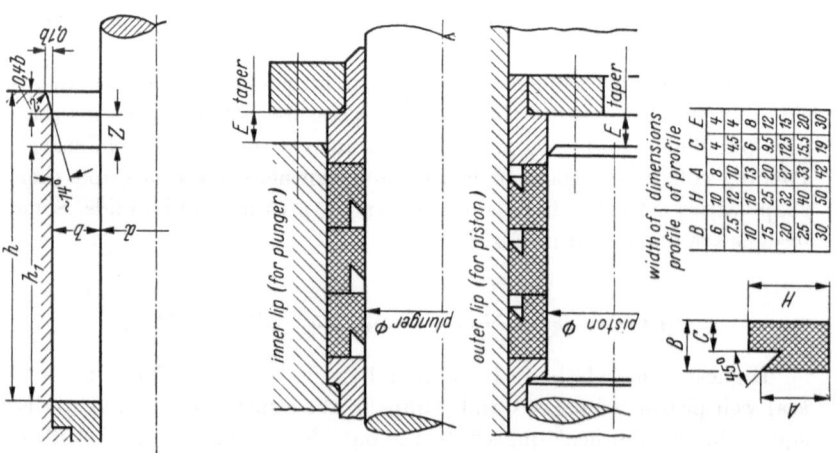

inner lip (for plunger)

outer lip (for piston)

width of profile / dimensions of profile

	B	H	A	C	E
	6	10	8	4	4
	2.5	12	10	4.5	4
	10	16	13	6	8
	7.5	25	20	9.5	12
	20	32	27	12.5	15
	25	40	33	15.5	20
	30	50	42	19	30

Table 24. Dimensions of Piston Rings and Spring Ring Packings (Goetzewerke)

d	D	$d + x$	a	b	h
40	60	40.1	49	50	5
70	86	70.1	77	78	5
100	122	100.1	110	111	6
150	181	150.1	165	166	8
200	240	200.1	219	220	10
250	299	250.1	274	275	13
300	353	300.2	326	327	13
400	465	400.2	432	433	16
500	580	500.2	540	541	20
600	699	600.2	650	651	25
700	820	700.3	760	762	30
800	920	800.3	860	862	30
900	1020	900.3	960	962	30

D	d	h_1	h_2	b
40	36	4	2	0.9
70	63.4	6	3	1.6
100	91.4	8	4	2
150	137.4	8	4	2.5
200	183.8	10	5	3
250	230.4	12	6	3.5
300	276.4	12	6	4
400	369.2	14	7	5
500	462	16	8	7
600	554.8	18	9	8
700	647.6	20	12	9
800	740.4	20	12	10
900	833.2	22	13	11

Table 23) so that it will not be necessary to go into further details[1]). Pistons of oil-hydraulic presses may also be sealed by cast-iron piston rings which are available in complete sets ready for installation (see Table 24).

---

[1]) MÜLLER, E.: Hydraulische Schmiedepressen und Kraftwasseranlagen, 2. Aufl. Berlin/Göttingen/Heidelberg: Springer 1952.

Such rings necessitate smoothly ground cylinder and plunger surfaces. Oil leakages being higher with cast-iron piston rings than with the usual soft packings, a leak oil chamber, from which the oil is passed to a storage tank, is provided in front of the spring ring packing close to the outlet of the piston shank.

c) Pipings and Accessories

Connection between an extrusion press and its pumping- or accumulator station is established by a pressure- and return piping with

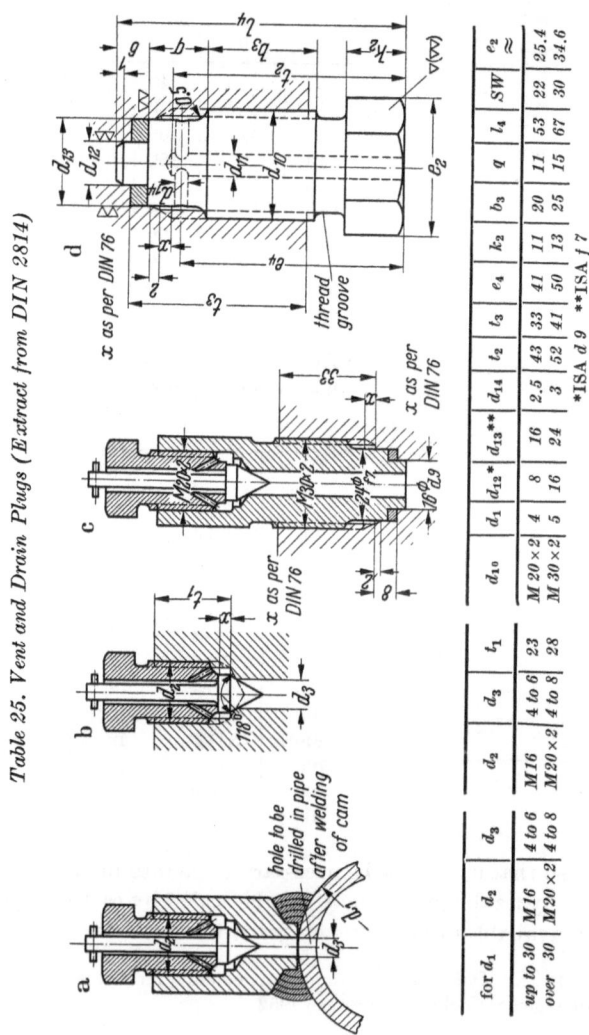

Table 25. Vent and Drain Plugs (Extract from DIN 2814)

for d_1	d_2	d_3		d_2	d_3	t_1		d_{10}	d_1	d_{12}*	d_{13}**	d_{14}	t_2	t_3	e_4	k_2	b_3	q	l_4	SW	e_2
up to 30	M16	4 to 6		M16	4 to 6	23		M 20×2	4	8	16	2.5	43	33	41	11	20	11	53	22	25.4
over 30	M20×2	4 to 8		M20×2	4 to 8	28		M 30×2	5	16	24	3	52	41	50	13	25	15	67	30	34.6

*ISA d 9 **ISA f 7

dimensions as per Tables 21 and 22. The pressure piping leads to the control gears and the return piping from the air-vessel to the storage tank. It must be noted that venting devices must be provided in all of the highest points in the circuit and drain devices in all of the lowest points to permit of draining the cylinders and pipings (see Table 25). All branch lines to the control gears must be provided with high-pressure shutoff devices and all drain pipes with three-way cocks so that it will not be necessary to drain the liquid from the system when removing the control valves. All of these accessories, i. e. pipes, flanged joints, manifolds, shutoff devices, venting- and draining devices, air-vessels and storage tanks, are known from other presses[1]). They are partly standardized and may be used for extrusion presses in the same design.

Chapter VIII

PRACTICAL EXAMPLES

a) Stresses of a Hydraulic Cylinder (Fig. 185)

Extrusion power $P = 1,400$ tons.

Hydraulic pressure $p_i = 315$ atm, respectively $p_i = -315$ atm under consideration of the sign for pressure in the calculation.

The cylinder is made of cast steel GS 52 with a yield point of $\sigma_s = 3,100$ kg/cm².

The extrusion power is transmitted via two integrally cast column sleeves which are connected with the cylinder through four walls being subjected to bending and tangential stresses.

The cylinder is subjected to a repetitive stress by the hydraulic pressure. Approx. 40 strokes are performed per hour so that after an assumed life of the press of 30 years, and two 8-hour shifts per working day, and 300 working days per year, the number of load alternations $n = 30 \times 300 \times 2 \times 8 \times 40 = 5,760,000$ is attained. Therefore, the fatigue limit has to be calculated as per WÖHLER's diagram.

In its upper part the cylinder represents a closed vessel, in its lower part an open vessel. The alternation is caused by the gradual transition of the axial stresses from the wall of the cylinder to the column sleeves.

The stresses are determined in the critical points and from these values the stress (comparison stress as per the theory of deformation) and compared with the permissible value.

1. Section a-a

The stresses resulting from the internal pressure $p_i = -315$ kg/cm² are according to BACH's formulas (see Table of Formulas 1, p. 169)

$$u = \frac{d_a}{d_i} = \frac{1,250}{775} = 1.615 \quad \text{and} \quad u^2 = 2.6$$

[1]) MÜLLER, E.: Hydraulische Schmiedepressen und Kraftwasseranlagen, 2. Aufl. Berlin/Göttingen/Heidelberg: Springer 1952.

inside the upper cylinder part:

$$\sigma_{t_i} = -p_i \frac{u^2 - 1}{u^2 - 1} = 315 \frac{3.6}{1.6} = \quad 709 \text{ kg/cm}^2$$

$$\sigma_{r_i} = p_i \qquad\qquad\qquad = -315 \text{ kg/cm}^2$$

$$\sigma_l = -p_i \frac{1}{u^2 - 1} = 315 \frac{1}{1.6} = \quad 197 \text{ kg/cm}^2$$

inside the lower cylinder part:

$$
\left.
\begin{aligned}
\sigma_{t_i} &= \quad 709 \text{ kg/cm}^2 \\
\sigma_{r_i} &= -315 \text{ kg/cm}^2 \\
\sigma_l &= 0 \text{ (open vessel)}
\end{aligned}
\right\}
$$ as in the upper part; however, decreasing in the packing space because of $p_i = 0$.

The bending stresses resulting from the column forces are in section a–a

$$\sigma_b = \pm \frac{Me}{J}$$

$$M = \frac{1,400,000}{2} \left(\frac{190}{2} - 16.45 \right) = 55,000,000 \text{ cm-kg} .$$

Distance e_1 of axis of gravity s–s: $e_1 = 107.2$ cm

hence $\quad e_2 = 184 - 107.2 = 76.8$ cm.

Moment of inertia J_s referred to axis of gravity s–s:

$$J_s = 38,220,000 \text{ cm}^4.$$

a) Compressive stress at lower cylinder part:

$$\sigma_{b1} = \frac{55,000,000}{38,220,000} (-107.2) = -154 \text{ kg/cm}^2$$

b) Tensile stress at transition from cylindrical part to bottom:

$$\sigma_{b2} = \frac{55,000\,000}{38,220,000} 76.8 = 111 \text{ kg/cm}^2$$

2. Section b-b:

The stresses resulting from the internal pressure $p_i = -315$ kg/cm², and $u = \frac{1,100}{775} = 1.42$ and $u^2 = 2.02$, are analogous to *1:*

inside the upper cylinder part:

$$\sigma_{t_i} = \quad 933 \text{ kg/cm}^2$$
$$\sigma_{r_i} = -315 \text{ kg/cm}^2$$
$$\sigma_l = \quad 309 \text{ kg/cm}^2$$

inside the lower cylinder part:

$$
\left.
\begin{aligned}
\sigma_{t_i} &= \quad 933 \text{ kg/cm}^2 \\
\sigma_{r_i} &= -315 \text{ kg/cm}^2
\end{aligned}
\right\}
$$ decreasing to 0.
$$\sigma_l = \quad 0$$

3. Section c-c:

Bending stress:

$$\sigma_b = \pm \frac{M}{W} ,$$

$$M = \frac{1,400,000}{4} \, 49 = 17,200,000 \text{ cm-kg,} \qquad W = \frac{12 \times 156^2}{6} = 48,670 \text{ cm}^3$$

$$\sigma_b = \pm \frac{17,200,000}{48,670} = \pm 353 \text{ kg/cm}^2 \,.$$

At the transition to the cylinder this stress is distributed approximately uniformly over the cylinder wall thickness. It reduces there to

$$\sigma_b = 353 \, \frac{12}{23.75} = 178 \text{ kg/cm}^2 \,.$$

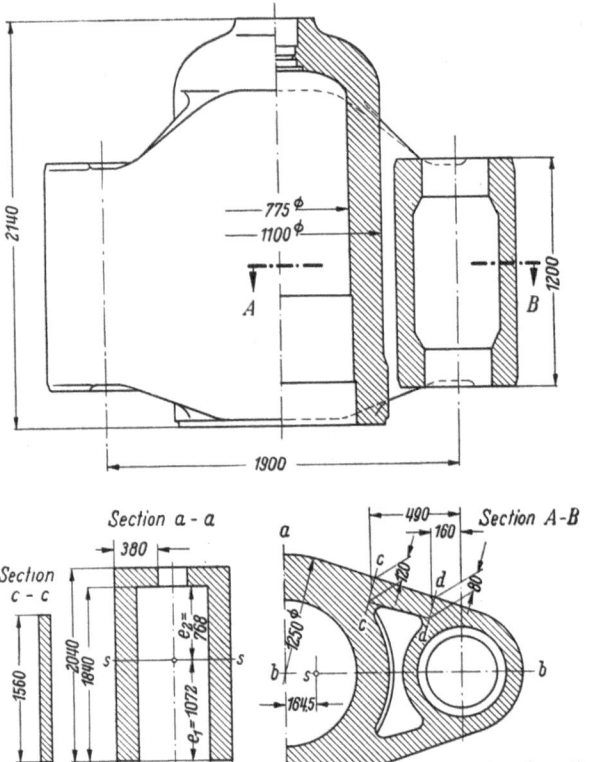

Fig. 185. Cast steel cylinder for a two-column press

In actual fact it will be smaller, because that part of the cylinder wall which surmounts the height of the ribbed wall, exerts a supporting effect.

Mean shearing stress in section c–c:

$$\tau_m = \frac{700,000}{156 \times 12 \times 2} = 187 \text{ kg/cm}^2 \,.$$

τ may be neglected when determining the comparison stress, because τ is equal to zero at the point of the maximum bending stress $\sigma_b = 353$ kg/cm² and, on the other hand, in τ_{max} there only exists a tangential tensile stress of

$$\sigma_{t_a} = -p_i \frac{2}{u^2 - 1} = 315 \frac{2}{1.02} = 618 \text{ kg/cm}^2 .$$

4. Section d-d:

$$\tau = \frac{700,000}{2 \times 8 \times 115} = 380 \text{ kg/cm}^2$$

and $\qquad \sigma = \pm \dfrac{700,000 \times 16 \times 6}{2 \times 8 \times 115^2} = \pm 318 \text{ kg/cm}^2 .$

Determination of Comparison Stresses

I. Section a-a:

The following total stresses exist at the transition from the cylindrical part to the bottom:

$$\sigma_{t_i} = 709 + 111 = \quad 820 \text{ kg/cm}^2$$
$$\sigma_{r_i} = -315 \text{ kg/cm}^2$$
$$\sigma_l = \quad 197 \text{ kg/cm}^2$$

According to the theory of deformation (see p. 169):

$$\sigma_v = \frac{1}{\sqrt{2}} \sqrt{(\sigma_t - \sigma_r)^2 + (\sigma_r - \sigma_l)^2 + (\sigma_l - \sigma_l)^2}$$

hence $\quad \sigma_v = 0.707 \sqrt{1,135^2 + 512^2 + 623^2} = 985 \text{ kg/cm}^2 .$

If surface coefficient $o_k = 1.5$ (for surface of casting), then a stress $\sigma_v = 1.5 \times 985 = 1,480 \text{ kg/cm}^2$ has to be taken into account.

At a tensile fatigue strength $\sigma_{sch} = 2,200 \text{ kg/cm}^2$, the safety factor is

$$S = \frac{2,200}{1,480} = 1.48 .$$

Closer examinations in the lower part are not necessary.

II. Section b-b:

At the transition from the cylindrical part to the bottom:

$$\sigma_{t_i} = \quad 933 \text{ kg/cm}^2$$
$$\sigma_{r_i} = -315 \text{ kg/cm}^2$$
$$\sigma_l = \quad 309 \text{ kg/cm}^2$$

and $\quad \sigma_v = 0.707 \sqrt{1,248^2 + 624^2 + 624^2} = \quad 1,081 \text{ kg/cm}^2$

Safety factor $S = \dfrac{2,200}{1.5 \times 1,081} = 1.36 .$

III. Section c-c:

Taking into account *1.* and *3.*:

$$\sigma_{t_i} = 709 + 178 = \quad 887 \text{ kg/cm}^2$$
$$\sigma_{r_i} = -315 \text{ kg/cm}^2$$
$$\sigma_l = \quad 197 \text{ kg/cm}^2$$

and $\quad \sigma_v = 0.707 \sqrt{1,202^2 + 512^2 + 690^2} = \quad 1,045 \text{ kg/cm}^2 .$

With surface coefficient $o_k = 1.5$, the stress to be taken into account is

$$\sigma_v = 1.5 \times 1,045 = 1,568 \text{ kg/cm}^2 \text{ and the}$$

$$\text{safety } S = \frac{2,200}{1,568} = 1.4.$$

IV. Section d-d:

The stresses calculated under *4.* on p. 242 are very small ones so that it will not be necessary to determine the comparison stresses.

The factor of safety 1 denotes the strength up to the end of the life assumed, if during each cycle the full load, as calculated, is applied. This is not the case in most of the hydraulic presses, as the water pressure occuring in the hydraulic cylinder is only such as required by the resistance of the respective material being extruded. The fact that the water pressure occurring is frequently lower than the nominal pressure, increases the life and the safety. Taking into account slight in-accuracies in the cast cylinder, as well as variations in the strength properties, theoretic safety factors are assumed at correspondingly higher values than 1. Same precautions are taken in case of instantaneously occurring loads.

b) Calculation of the Four Colums of a 2,500-Ton Tube and Rod Extrusion Press (Fig. 186)

Due to the high number of load alternations the fatigue limit (tensile fatigue strength) has to be calculated. The column is additionally subjected to bending by its own weight; however, the bending stress occurring is so little that it may be neglected.

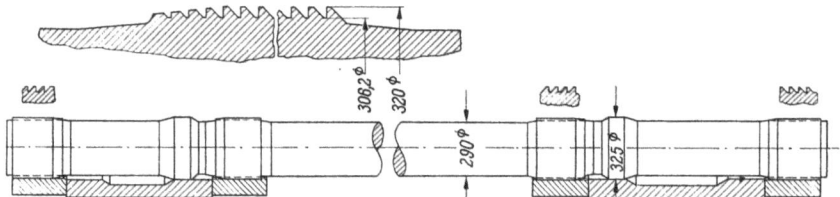

Fig. 186. Press column with saw-tooth thread

The nominal tensile stress in the cross-section F of the core of the thread is with

$$P_s = \frac{P}{4} = 625 \text{ tons and } F = 736 \text{ cm}^2: \quad \sigma_n = \frac{P_s}{F} = \frac{625,000}{736} = 850 \text{ kg/cm}^2.$$

The thread is in saw-tooth form as per DIN 2781 (see Table 19) with approximately the same fatigue-stress concentration factor as that of Whitworth threads; consequently $\beta_k = 2.2 - 3.2$ for medium carbon steel. Let $\beta_k = 2.7$, then the maximum stress at the notch is

$$\sigma_{\max} = \beta_k \sigma_n = 2.7 \times 850 = 2,300 \text{ kg/cm}^2.$$

The columns are made of steel C 35 having a yield point $\sigma_s = 3,200 \text{ kg/cm}^2$ and approximately equal fatigue limit, consequently $\sigma_s = \sigma_{sch}$. Then the safety factor

16 E

against fatigue fracture is

$$S = \frac{\sigma_{sch}}{\sigma_{max}} = \frac{3,200}{2,300} = 1.4 \,.$$

This factor of safety is increased by prestressing the column length between two nuts. Thus the stress will become a predominantly static one with overlapping repetitive stresses. Frequently engineering and technological steps are taken to reduce β_k. The scale effect – which reduces the fatigue strength and which has not been clarified sufficiently so far – is thus catered for, too. In order to ensure a better distribution of the total load over the individual pitches, the latter are provided with a slight taper starting from the stressed face of the nut.

It has still to be mentioned that the middle part of the column with some of the pitches, which cannot be prestressed, is subjected to a full repetitive stress; however, no forces are transmitted from these pitches to the nut. The fatigue-stress concentration factor existing here is therefore the considerably lower one of the notched bar.

c) Calculation of a Cable Sheathing Press

A cable of the following dimensions is to be extruded from one charge of the container of a cable sheathing press: 55 mm outside diameter; 1.8 mm wall; length $l = 110$ m; emerging speed $v_a = 20$ m/min; without stop mark.

Taking into account the extrusion of lead alloys, the extrusion pressure $p = P/F = 5,000$ kg/cm² has to be assumed, this also covering the mechanical and hydraulic efficiency of the press.

The extrusion time is $t = \dfrac{l}{v_a} = \dfrac{110}{20} = 5.5$ minutes.

With a solidifying period of 7.5 minutes and an idle time of 2 minutes required for the operation of the press, one extrusion cycle is performed in 15 minutes so that 4 extrusions are performed per hour.

The effective weight G_N of one charge is determined from the given dimensions of the sheath:

$$G_N = \left(\frac{\pi d_a^2}{4} - \frac{\pi d_i^2}{4} \right) l\gamma = f l \gamma \,,$$

$$G_N = (23.76 - 20.75) \times 11,000 \, \frac{11.1}{1,000} = 366 \, \text{kg} \,.$$

Taking into account about 3% to cover for losses due to the formation of a shell on the extrusion ram, the weight of the charge is:

$$G_F = \frac{G_N}{0.97} = 378 \, \text{kg} \,.$$

With four extrusions per hour, the weight of the lead extruded per hour is

$$G_h = 4 G_N = 1,464 \, \text{kg} \,.$$

The diameter of the extrusion ram D – assuming an appropriate ratio of D to stroke H – results from the filling volume V_F.

If $H = 5 D$, then $V_F = \dfrac{G_F}{\gamma} = F H = \dfrac{\pi D^2}{4} H = \dfrac{5}{4} \pi D^3 \,,$

hence $D = \sqrt[3]{\dfrac{0.8 \, G_F}{\pi \gamma}} = \sqrt[3]{\dfrac{0.8 \times 378}{\pi \times 11.1}} \cong 2.05 \, \text{dm} = 205 \, \text{mm}$

and the power stroke $H = 5 D = 1,025$ mm.

From the face of the ram $F = \dfrac{\pi D^2}{4} = 330$ cm² and the cross-sectional area of the extruded sheath $f = 3.01$ cm² results the extrusion ratio $\varphi = \dfrac{F}{f} = \dfrac{330}{3.01}$ $= 110$ and hence the ram speed $v = \dfrac{v_a}{\varphi} = \dfrac{20}{110} = 0.182$ m/min $= 3.04$ mm/sec.

The ram power

$$P = \frac{\pi D^2}{4}\, p, \text{ if } p = 5{,}000 \text{ kg/cm}^2 = 5 \text{ ton/cm}^2$$

$$P = 330 \times 5 = 1{,}650 \text{ tons.}$$

Choosing an operating water pressure $p_w = 315$ atm, then the diameter of the main plunger D_1 results from $\dfrac{\pi D_1{}^2}{4}\, p_w = F_1\, p_w = P$.

Therefore

$$F_1 = \frac{P}{p_w} = \frac{1{,}650{,}000}{315} \cong 5{,}250 \text{ cm}^2$$

and hence $D_1 = 820$ mm.

From the consumption of pressure water $Q = F_1 v = 52.5 \times 1.82 = 95.5$ l/min is calculated the capacity of the pump

$$N = \frac{Q\,p_w}{360} = \frac{95.5 \times 315}{360} \cong 84 \text{ H. P.}$$

An empirical value $P_R = 0.1\, P$ may be assumed as the maximum force required for the return stroke. Hence $P_R = 165$ tons and the area of the pullback plunger $F_R = 0.1\, F_1 = 525$ cm².

From the pump delivery $Q = 95.5$ l/min results a return speed $v_R = v F_1/F_R$ $= 3.04 \times 10 \cong 30$ mm/sec.

d) Calculation of a Tube and Rod Extrusion Press

A tube and rod extrusion press is to be capable to extrude copper and brass tubes of a maximum outside diameter $d_a = 120$ mm, a 5 mm wall and in minimum lengths of 8 m, over a mandrel that is held in the die, under the following conditions:

The maximum theoretical ram pressure is to be $p = P/F = 7{,}400$ kg/cm², the piercer pressure $p_L = P_L/F_L = 3{,}200$ kg/cm². Mechanical and hydraulic losses are included.

The extrusion ratio is to be at least $\varphi = F/f = 15$, where $F =$ cross-sectional area of dummy-block and $f =$ cross-sectional area of the extruded tube.

The maximum billet length in rod extrusion is to be $l = 3.6\, d$ ($d =$ billet diameter).

A maximum ram speed $v = 100$ mm/sec is required.

Operating water pressure $p_w = 200$ atm;

Cross-sectional area of tube $f = \dfrac{\pi}{4}(d_a{}^2 - d_i{}^2) = 18.06$ cm² ;

Cross-sectional area of dummy-block $F = \varphi f = 15 \times 18.06 = 271$ cm²;

Power of extrusion ram $P = Fp = 271 \times 7{,}400 = 2{,}000{,}000$ kg $= 2{,}000$ tons;

Pullback power of ram $P_R \cong 0.1\, P = 200$ tons;

16*

Area of pullback plunger $F_R = P_R/p_w = 1,000$ cm^2 (chosen: 2 pullback plungers

each $D_R = 250$ mm, total $F_R = 982$ cm^2);

Piercer force $P_L = \dfrac{\pi\, d_i^{\,2}}{4}\, p_L = 95 \times 3,200 = 304,000$ kg $\cong 300$ tons ;

Area of piercer plunger $F_L = P_L/p_w = 1,500$ cm^2;

Diameter of piercer plunger chosen $D_L = 440$ mm with $F_L = 1,521$ cm^2;

Piercer mandrel pullback force $P_{RL} \cong 0.15\, P_L = 45$ tons;

Area of piercer mandrel pullback plunger $F_{RL} = P_{RL}/p_w = 225$ cm^2 (chosen:

2 plungers each $D_{RL} = 120$ mm, total $F_{RL} = 226$ cm^2);

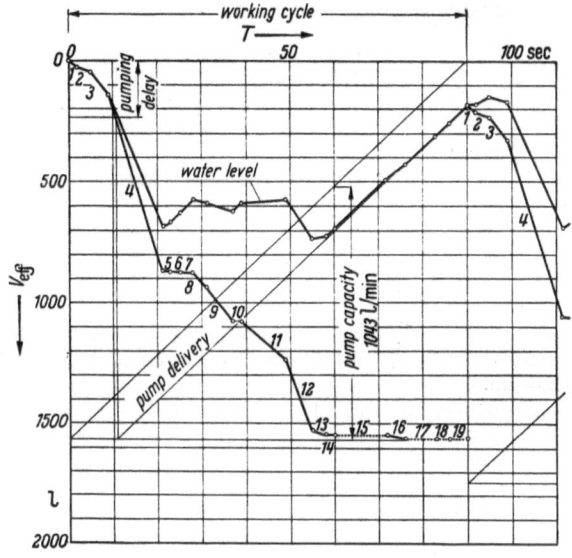

Fig. 187. Water intake and pump delivery of a tube and rod extrusion press during working
cycle as per Table 26

The diameter of the dummy-block is given by

$$\frac{\pi\, d_1^{\,2}}{4} = F + \frac{\pi\, d_i^{\,2}}{4} = 271 + 95 = 366 \text{ cm}^2, \text{ resulting in } d_1 = 216 \text{ mm} ;$$

Diameter of container for extrusion "with shell" chosen $d_2 = 220$ mm;

Billet diameter chosen $d = 215$ mm;

Initial billet length $l = 485$ mm, corresponding to a tube length of about 8.5 m,
when assuming about 12% of volume of billet for discard and shell;

Length of billet upset in container $l_1 \cong 460$ mm;

Length of billet upset and pierced $l_2 \cong 600$ mm;

Length of container for tube extrusion $L = l_2 + H + t = 735$ mm, if

Thickness of dummy-block $H = 105$ mm and

Projection of die in container $t = 30$ mm.

Table 26. Consumption of Pressure Water and Cycle Time

	Operation	Pressure Water Consumed by (P. = Plunger Piston D. = Disk Piston)	Eff. Area cm²	Eff. Stroke mm	Consumption litres	Time Incl. Operating Time seconds
1	Upsetting of billet	main P.	10,000	25	25	2
2	Short return	2 press pullback P.	982	150	15	3
3	Piercing	piercer P.	1,521	485	74	4
4	Extrusion	main P.	10,000	580	580	12
5	Short return	2 press pullback P. 2 piercer pullback P.	982 226	30 70	3 2	} 2
6	Container stripping	2 container shifting D.	57	20	0.1	2
7	Locking slide up	const. pullback P.	67	450	3	3
8	Ejection of dummy-block and discard	main P.	10,000	180	180	3
9a	Die-carrier out	die-carrier shifting P.	38	2,880	11	} 6
9b	Return of main and piercer P.	2 press pullback P. 2 piercer pullback P.	982 226	1,000 1,000	100 23	
10	Ejector disk in	—	—	—	—	2
11a	Ejection of shell	—	—	—	—	} 10
11b	Shearing of discard and return of shear	shear P. shear pullback P.	616 30	420 420	26 1	
12a	Return of main and piercer P.	2 press pullback P. 2 piercer pullback P.	982 226	1,935 1,935	190 44	} 6
12b	Die-carrier in	die-carrier shifting P.	38	2,880	11	
13	Locking slide down	locking slide P.	20	450	1	3
14	Container sealing	container shifting D.	57	20	0.1	2
15	Mandrel out and cooling	—	—	—	—	12
16	Return of mandrel	2 piercer pullback P.	226	630	14	4
17	Charging of billet from furnace to press	—	—	—	—	7
18	Dummy-block in	—	—	—	—	3
19	Advance of ram and mandrel to billet	—	—	—	—	4
				$V_{theor.} = 1,303\ 1$		$T = 90\ \text{sec.}$

The effective area of pressure of the main plunger $F_P = \dfrac{P}{p_w} = \dfrac{2,000,000}{200}$
$= 10,000 \text{ cm}^2$ consists of the annular area between the diameter D of the main plunger and the diameter D_1 of the plunger extension. Assuming the diameter D_2 of the mandrel bar to transmit the piercer force $P_L = 300$ tons at $D_2 = 280$ mm and the wall thickness of the plunger extension at $s = 95$ mm, then $D_1 = D_2 + 2s = 280 + 2 \times 95 = 470$ mm and with $F_P = 10,000 = \pi/4\,(D^2 - D_1{}^2)$ is given the diameter of the main plunger $D \cong 1,225$ mm.

Starting from the longest billet to be loaded with the dummy-block between container and retracted ram, the stroke of the main ram is $H_P = L + 2H + l_3 + 50$, where the maximum billet length for rod extrusion $l_3 = 3.6\,d = 3.6 \times 215 + 770$ mm and the container length $L = l_3 + H + t = 770 + 105 + 30 = 905$ mm, so that the stroke $H_P = 905 + 2 \times 105 + 770 + 50 = 1,935$ mm.

The forces and strokes determined in the draft, are shown in Table 26 and Fig. 187, which have been tabulated for the calculation of the power water station.

With the cycle time $T = 90$ seconds, as per Table 26, the number of extrusions per hour is $n = \dfrac{3,600}{T} = \dfrac{3,600}{90} = 40.$

The pressure water consumption in one cycle, under due consideration of the compression of the water in the piping and cylinders as well as inclusions of air and leaks, is calculated at

$$V_{\text{eff}} = 1.2\ V_{\text{theor}} = 1.2 \times 1,303 = 1,564 \text{ litres.}$$

Hence the pump delivery $Q = V_{\text{eff}}\,\dfrac{60}{T} = 1,564\,\dfrac{60}{90} = 1,043$ l/min and the motor power $N = \dfrac{Q\,p_w}{360} = \dfrac{1,043 \times 200}{360} = 580$ H. P.

The pressure water contents of the accumulator is calculated so as to allow for a cycle, which has been commenced, to be completed in case of breakdown of a pump caused, for example, by failure of current. Adding moreover a safety allowance of 10%, then the pressure water contents $I = 1.1 \times 1,564 = 1,720$ litres.

The compressed-air contents of the air-bottles is generally chosen so that the pressure drop, if the water volume V_{eff} is withdrawn, will be about 12%.

With polytropic expansion of air $p_1 V_1{}^n = p_2 V_2{}^n$.

With an exponent n, which may be assumed in the pressure range of 200 atm at $n = 1.3$, the volume of air required is $V_L = 9.7\ V_{\text{eff}} \cong 15,200$ litres.

e) Dimensions and Shrinkage Allowances of a 3-Part Container (Fig. 188)

Mean diameter of the inner bore	$d_0 = 230$ mm
Press power	$P = 2,500$ tons
Maximum extrusion pressure in tube extrusion (with negative sign in the calculation)	$p = -90 \text{ kg/mm}^2$
Maximum radial pressure	$p_i = -90 \times 0.7 = -63 \text{ kg/cm}^2$ assumed
Outside diameter of liner	$d_1 = 322$ mm $= 1.4\,d_0$ chosen
Outside diameter of sleeve	$d_2 = 515$ mm $= 1.6\,d_1$ chosen
Outside diameter of jacket	$d_3 = 1,000$ mm $= 1.94\,d_2$ chosen.

The compressive prestress caused by the contraction strain, is to be chosen so that no tangential tensile stresses will occur in the liner during operation.

The maximum operating temperature of the container is 550 °C.

With the help of BACH's Formulas the stresses σ_t and σ_r are determined from the pressures p_i, p_{s2} and p_{s1} (see Table of Formulas 1, p. 169, and Fig. 189).
Stresses from $p_i = -63$:

$$\sigma_{t0} = -p_i \frac{u^2+1}{u^2-1} = 63 \frac{19.9}{17.9} = 70 \qquad\qquad \sigma_{r0} = p_i \qquad\qquad\qquad = -63$$

$$\sigma_{t1} = -p_i \frac{U_2{}^2+1}{u^2-1} = 63 \frac{10.64}{17.9} = 37.4 \qquad \sigma_{r1} = p_i \frac{U_2{}^2-1}{u^2-1} = -63 \frac{8.64}{17.92} = -30.4$$

$$\sigma_{t2} = -p_i \frac{u_2{}^2+1}{u^2-1} = 63 \frac{4.77}{17.9} = 16.8 \qquad \sigma_{r2} = p_i \frac{u_2{}^2-1}{u^2-1} = -63 \frac{2.77}{17.92} = -9.8$$

$$\sigma_{t3} = -p_i \frac{2}{u^2-1} = 63 \frac{2}{17.9} = 7 \qquad\qquad \sigma_{r3} = \qquad\qquad\qquad\quad = 0$$

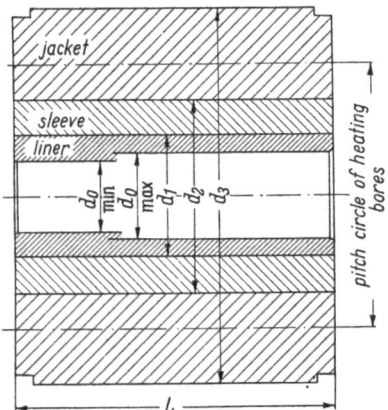

Fig. 188. Three-part container

Stress $\sigma_{t0} = 70$, caused by $p_i = -63$, is to be balanced by an opposed, equal-sized prestress $\sigma_{t0} = -70$. Consequently, by shrinking stresses p_{s2} and p_{s1} must be produced a radial pressure, which produces the prestress $\sigma_{t0} = -70$. In point 1 (Fig. 189) this radial pressure has a value of

$$\sigma_{r1} = \sigma_{t0} \frac{u_0{}^2-1}{2\,u_0{}^2} = -70 \frac{0.96}{3.92} = -17.1 \,.$$

It is composed of p_{s1} and the radial pressure caused by p_{s2} in point 1.
The determination of p_{s2} is based on a permissible stress σ_{t2} which occurs in the jacket during operation. Let $\sigma_{t2} = 45$, then in connection with σ_{r2} and p_{s2} a σ_{v2} occurs in the jacket which is not likely to exceed the permissible value. As a $\sigma_{t2} = 16.8$ which results from $p_i = -63$, exists already in point 2, there must be produced from p_{s2} a $\sigma_{t2} = 45 - 16.8 = 28.2$, from which may be determined p_{s2}.

$$\sigma_{t2} = 28.8 \text{ requires } p_{s2} = -\sigma_{t2} \frac{u_2{}^2-1}{u_2{}^2+1} = -28.2 \frac{2.77}{4.77} = -16.4 \,.$$

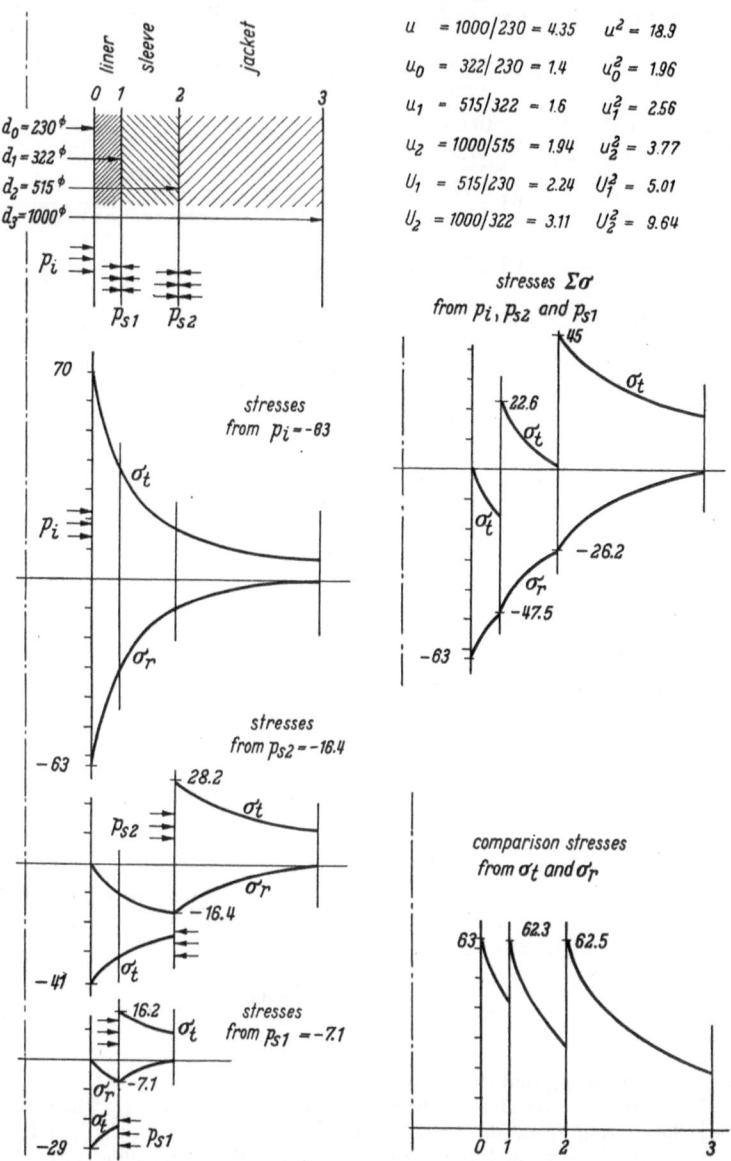

Fig. 189. Stresses in a three-part container

This $p_{s2} = -16.4$ produces in liner and sleeve:

$$\sigma_{t0} = p_{s2}\frac{2U_1^2}{U_1^2-1} = -16.4\frac{10.04}{4.01} = -41 \qquad \sigma_{r0} = \qquad = 0$$

$$\sigma_{t1} = p_{s2}\frac{U_1^2+u_1^2}{U_1^2-1} = -16.4\frac{7.58}{4.01} = -31 \qquad \sigma_{r1} = p_{s2}\frac{U_1^2-u_1^2}{U_1^2-1} = -16.4\frac{2.46}{4.02} = -10$$

$$\sigma_{t2} = p_{s2}\frac{U_1^2+1}{U_1^2-1} = -16.4\frac{6.02}{4.01} = -25 \qquad \sigma_{r2} = \qquad = -16.4$$

in the jacket:

$$\sigma_{t2} = -p_{s2}\frac{u_2^2-1}{u_2^2-1} = 16.4\frac{4.77}{2.77} = 28.2 \qquad \sigma_{r2} = \qquad = -16.4$$

$$\sigma_{t3} = -p_{s2}\frac{2}{u_2^2-1} = 16.4\frac{2}{2.77} = 11.8 \qquad \sigma_{r3} = \qquad = 0$$

After shrink-assembly of sleeve and jacket a radial pressure is to occur in point 1, which has been determined at $\sigma_{r1} = -17.1$. From p_{s2} occurs already a radial pressure $\sigma_{r1} = -10$ in point 1. Consequently, a further radial pressure $p_{s1} = -17.1 + 10 = -7.1$ is required which must be produced solely by shrinking the sleeve over the liner prior to shrink assembling the jacket.

$p_{s1} = -7.1$ produces
in the liner:

$$\sigma_{t0} = p_{s1}\frac{2u_0^2}{u_0^2-1} = -7.1\frac{3.92}{0.96} = -29 \qquad \sigma_{r0} = \qquad = 0$$

$$\sigma_{t1} = p_{s1}\frac{u_0^2+1}{u_0^2-1} = -7.1\frac{2.96}{0.96} = -21.9 \qquad \sigma_{r1} = p_{s1} \qquad = -7.1$$

in the sleeve:

$$\sigma_{t1} = -p_{s1}\frac{u_1^2+1}{u_1^2-1} = 7.1\frac{3.56}{1.56} = 16.2 \qquad \sigma_{r1} = p_{s1} \qquad = -7.1$$

$$\sigma_{t2} = -p_{s1}\frac{2}{u_1^2-1} = 7.1\frac{2}{1.56} = 9.1 \qquad \sigma_{r2} = \qquad = 0$$

The similar stresses resulting from p_i, p_{s2} and p_{s1} overlap during operation and are therefore to be added algebraically.

Addition:

stress	due to	0 liner		1 sleeve		2 jacket	3
σ_t	$p_i = -63$	$+70$	$+37.4$	$+37.4$	$+16.8$	$+16.8$	$+7$
	$p_{s2} = -16.4$	-41	-31	-31	-25	$+28.2$	$+11.8$
	$p_{s1} = -7.1$	-29	-21.9	$+16.2$	$+9.1$	$-$	$-$
	$\Sigma\sigma_t$	0	-15.5	$+22.6$	$+0.9$	$+45$	$+18.8$

stress	due to	0 liner	1	sleeve	2	jacket	3
	$p\ \ = -63$	-63	-30.4	-30.4	-9.8	-9.8	0
σ_r	$p_{s2} = -16.4$	0	-10	-10	-16.4	-16.4	0
	$p_{s1} = -\ 7.1$	0	-7.1	-7.1	0	—	—
	$\Sigma\,\sigma_r$	-63	-47.5	-47.5	-26.2	-26.2	0

From the summation stresses determined in points 0, 1, 2 and 3, the comparison stress in each of these points may be established. In order to avoid permanent sets, the comparison stress must not exceed the yield strength at elevated temperature.

According to the theory of deformation, the comparison stress is

$$\sigma_c = \frac{1}{\sqrt{2}}\,\sqrt{(\sigma_t - \sigma_r)^2 + (\sigma_r - \sigma_l)^2 + (\sigma_l-\sigma_t)^2}\,.$$

In the present case an *open* container is concerned, which on account of p_i is not subjected to axial stresses σ_l. In such case is assumed $\sigma_l = 0$, so that the formula for σ_v is simplified to

$$\sigma_c = \sqrt{\sigma_t{}^2 + \sigma_r{}^2 - \sigma_t\,\sigma_r}\,.$$

Hence, inserting the determined values for σ_t and σ_r, the comparison stresses are:

liner $\begin{cases} \sigma_{v0} = 63 \\ \sigma_{v1} = 41.9 \end{cases}$ sleeve $\begin{cases} \sigma_{v1} = 62.3 \\ \sigma_{v2} = 27.1 \end{cases}$ jacket $\begin{cases} \sigma_{v2} = 62.5 \\ \sigma_{v3} = 18.8 \end{cases}$.

The last curves in the graphs Fig. 189 illustrate the comparison stresses. They show that the loads in the three parts of the container are equal to each other. This means that with the diameters and contractions chosen, providing equal quality of materials and equal heating of all of the parts, equal efficiency is attained.

In practice, however, variations occur due to the degree of accuracy in machining and the unavoidable temperature gradient caused by the overheating of the liners.

From the fact that containers have proved in practice, which with regard to materials and shrinkage allowances correspond to the values upon which the calculation is based, it may be concluded that certain favorable influences are active, too: The assumption, for example, that radial stresses do not occur, is not correct. During extrusion, on the one hand, the container is forced against the die and, on the other hand, compressive forces are transmitted to the container by the wall friction of the flowing billet. If these axial compressive forces are inserted into the present calculation, the comparison stress may be reduced considerably.

According to the theory of deformation, the optimum axial stress from which results at given values σ_t and σ_r the smallest comparison stress, is:

$$\sigma_{l\,opt} = \frac{\sigma_t + \sigma_r}{2}\,.$$

The calculation being based on the maximum internal pressure, which will hardly ever occur in its full value and would presuppose the concurrence of the most unfavorable circumstances, i. e. difficultly extrudable billet material at maximum billet length, it may be assumed that the hazard of maximum stress will be present in extraordinarily rare cases only. Therefore, the stress may be calculated within close limits of the yield point without risking to impair the life of the container considerably.

Determination of Shrinkage Allowances. The inner bore of the part to be shrunk on is given the nominal diameter. After finish-machining of this bore, the actual dimension, which must be well in the permissible tolerances, is measured and from this value is determined the outside diameter of the inner part which must be larger by the shrinkage allowance s.

The shrinkage allowance s, which is to produce the calculated contraction stress p_s, is given by:

$$s = \frac{d}{E_t} (\sigma_{t_g} - \sigma_{t_k}) \quad \text{(see p. 170)}.$$

(see p. 170)

In Point 1:

$d = 322$

$\sigma_{t_g} = +16.2$ (sleeve)

$\sigma_{t_k} = -21.9$ (liner)

$E_t = 16,000$

and hence

$$s_1 = \frac{322}{16,000} (16.2 + 21.9) = 0.77 \text{ mm};$$

In Point 2:

$d = 515$

$\sigma_{t_g} = +28.2$ (jacket)

$\sigma_{t_k} = -25$ (sleeve)

$E_t = 16,000$

and hence

$$s_2 = \frac{515}{16,000} (28.2 + 25) = 1.77 \text{ mm}.$$

INDEX